T0059362

Also by Kevin Cook

Tommy's Honor

Driven

Titanic Thompson

The Last Headbangers

Flip

Kitty Genovese

The Dad Report

Electric October

Ten Innings at Wrigley

THE
BURNING
BLUE

THE BURNING BLUE

The Untold Story of Christa McAuliffe and NASA's Challenger Disaster

Kevin Cook

HENRY HOLT AND COMPANY

NEW YORK

Henry Holt and Company
Publishers since 1866
120 Broadway
New York, New York 10271
www.henryholt.com

Henry Holt® and Ⓗ® are registered trademarks of Macmillan Publishing
Group, LLC.

Copyright © 2021 by Kevin Cook
All rights reserved.

Library of Congress Cataloging-in-Publication Data

Names: Cook, Kevin, 1956– author.
Title: The burning blue : the untold story of Christa McAuliffe and
  NASA's Challenger disaster / Kevin Cook.
Description: First edition. | New York, New York : Henry Holt and
  Company, 2021. | Includes bibliographical references and index.
Identifiers: LCCN 2020052641 (print) | LCCN 2020052642 (ebook) |
  ISBN 9781250755551 (hardcover) | ISBN 9781250755568 (ebook)
Subjects: LCSH: Challenger (Spacecraft)—Accidents. | McAuliffe,
  Christa, 1948–1986—Friends and associates. | Space flight training—
  United States. | United States. National Aeronautics and Space
  Administration—Management.
Classification: LCC TL867.C6529 2021 (print) | LCC TL867 (ebook) |
  DDC 363.12/4—dc23
LC record available at https://lccn.loc.gov/2020052641
LC ebook record available at https://lccn.loc.gov/2020052642

Our books may be purchased in bulk for promotional, educational, or
business use. Please contact your local bookseller or the Macmillan
Corporate and Premium Sales Department at (800) 221-7945, extension
5442, or by e-mail at MacmillanSpecialMarkets@macmillan.com.

First Edition 2021

Designed by Meryl Sussman Levavi

Printed in the United States of America

1   3   5   7   9   10   8   6   4   2

To the memory of Dick Scobee, Michael Smith,
Ellison Onizuka, Judith Resnik, Ronald McNair,
Gregory Jarvis, and Christa McAuliffe

Up, up the long, delirious, burning blue
I've topped the wind-swept heights with easy grace
Where never lark nor ever eagle flew—
And, while with silent lifting mind I've trod
The high untrespassed sanctity of space,
Put out my hand, and touched the face of God.

—from the poem "High Flight,"
by John Gillespie Magee

PROLOGUE

T HEY WERE ALL ON EDGE. THEY'D SPENT THE DAY BEFORE THE same way, strapped to their seats in launch position for five hours, lying on their backs looking up at their knees. After that—the third scrub in three days—a friend asked Christa McAuliffe how it felt to be cooped up for so long.

"Put on a motorcycle helmet," she said. "Lie on the floor with your legs up on a bed. You can't read, you can't watch television. You're strapped down, with oxygen lines and wires coming out of your suit. Lie there for five hours and you'll know how it feels."

Today's delay was an hour and counting. Overnight, record-setting cold had frosted crops in the orange groves and strawberry fields near Cape Canaveral. Now, on the coldest morning in twenty years, the crew of the space shuttle *Challenger* waited while NASA workers used broomsticks to knock icicles off the shuttle.

It was chilly on the flight deck—*Challenger*'s cockpit. Mission specialist Ellison Onizuka said his nose was frozen. Mission specialist Judith Resnik claimed she had it worse: "My butt is dead." Below them, on the windowless middeck, payload specialist McAuliffe dozed through the latest delay. Then the radio crackled. "This is NASA tower. We are planning to come out of this hold on time"—the best possible news. Commander Dick Scobee radioed back: "Roger. That's great."

Six flight-deck windows gave Scobee a panoramic view of cold blue sky. "Well, y'all on the middeck," he said, sounding like an airline pilot, "it's clear blue out today."

Mike Smith, sitting in the pilot's seat to Scobee's right, flicked switches that activated the ship's auxiliary power units. "APUs coming on," he said.

Scobee checked his instrument panel. "Pressure on all three APUs." Next, he confirmed that each crew member was sealing his or her flight helmet. "Visors coming down." At T minus two minutes, he thumbed the intercom to Smith, Resnik, Onizuka, McAuliffe, mission specialist Ron McNair, and payload specialist Greg Jarvis. "Welcome to space, guys," he said.

At T minus 1:47, a robot arm pulled back the hood of the rust-colored fuel tank that dwarfed the shuttle clamped to its back. "There goes the beanie cap," Scobee said.

"Doesn't it go the other way?" Onizuka asked. He was joking.

"God, I hope not . . . thirty seconds."

At T minus sixteen seconds, water cannons sprayed three hundred thousand gallons of water into the trench below "the stack"—the 184-foot-tall contraption made up of the shuttle, its giant fuel tank, and a pair of rocket boosters. The water would muffle the thunder of launch, keeping sound waves coming off the engines from pounding upward through the stack's fiery exhaust and damaging the shuttle.

At T minus ten, loudspeakers at the launchpad and the bleachers at Kennedy Space Center, three miles away, carried the countdown. "Nine, eight . . ." At T minus six, the crew felt the cabin shake as the first of the shuttle's three onboard engines came to life, followed by two more booms from the second and third. The engines' thrust made the towering stack sway sideways by almost two feet—the so-called twang effect a queasy rider like McAuliffe could feel in her stomach.

Scobee said, "Three at a hundred." All three engines were at full power. In the next instant, guided by the shuttle's computers, the stack returned to vertical and the solid rocket boosters fired. The silo-shaped SRBs, each weighing 1.3 million pounds, would provide 80 percent of the force required to thrust the shuttle, crew, and cargo into orbit. They were the most powerful rockets ever built. Astronauts had a saying: "Once those SRBs get lit, the stack's going *somewhere*. You just hope it's the right direction."

Eight massive bolts held the stack to the launchpad. Each bolt was wired to an explosive charge. At the instant the SRBs ignited—11:38:00:01 a.m. Eastern time on January 28, 1986—the bolts detonated, and the stack began to rise. Clearing the launch platform in clouds of fire and white exhaust, it took eight seconds to reach a hundred miles per hour. Within a minute it was moving fifteen times as fast, the speed of a rifle bullet. The crew held on while the two-billion-dollar shuttle shook and groaned like a rustbucket freighter in a typhoon. No astronaut-training simulator came close to matching the bone-rattling racket of an actual launch.

The windows shook. Shock waves sent shivers through the cabin walls, through the astronauts' steel seats and the fillings in their teeth, pressing them earthward at three Gs, enough to make a 128-pound social studies teacher feel like she weighed 384. But this was the moment she'd been dreaming of for the past year. Astronauts had told her it would be loud, and it was—loud and scary—but now they were less than two minutes from orbit.

Commander Scobee and pilot Smith talked to Mission Control. Mission specialists Resnik, Onizuka, and McNair had duties of their own. McAuliffe, America's "Teacher in Space," had nothing to do but hold on.

1

SHARON CHRISTA CORRIGAN WAS A SEVENTH GRADER IN Framingham, Massachusetts, on May 5, 1961, the day Alan Shepard became the first American to fly into space. Christa, as everyone called her, joined classmates to watch the launch on a portable TV in the school cafeteria. The grainy black-and-white screen showed Shepard in his cramped capsule atop an eighty-foot Mercury-Redstone rocket that could launch him skyward or blow him to bits. After three hours of glitches and delays, Shepard was pissed. He radioed Launch Control: "Fix your little problems," he said, "and *light this candle.*"

Christa Corrigan grew up with the space program. Her favorite TV hero was Superman, the man who flew faster than rockets. Her political hero was President Kennedy, who announced that the United States would put a man on the moon before the decade was out. As a schoolgirl she followed Shepard's suborbital flight and safe landing, John Glenn's 1962 orbits of Earth, and the rest of the Mercury and Gemini programs. She thought it would be neat to be an astronaut, but Christa was a practical person. America was out to put a man on the moon, not a robot, a monkey, or a woman. She wouldn't have made much of an astronaut anyway, a chubby Girl Scout with no knack for science or math who got sick to her stomach on carnival rides.

She had barely survived to go to school in the first place. As an infant, she spent her first few weeks fighting a gastrointestinal illness, wailing and wasting away at Boston Children's Hospital while her parents held her little hands and prayed. Doctors kept the baby alive by poking tubes into her arms and scalp, feeding her a mixture of glucose and water until a new antibiotic, Aureomycin, saved her life. After that she kept charging at life as if life was a gift. As a toddler, she rode her tricycle into traffic on busy Columbia Street. Three-year-old Christa pedaled for all she was worth, cars zipping by in both directions. The family dog, a mutt named Teddy, took off after her. Teddy yapped and ran circles around the little girl on the trike until traffic stopped. Grace Corrigan corralled her daughter and led her home, giving thanks to Saint Christopher, the patron saint of travelers, for whom the girl was named.

As an overachieving high schooler, "there was a special vibrancy to her," recalled one of the nuns who taught her at Framingham's Marian High School. While babysitting four younger siblings, taking piano and guitar lessons, and working on weekends at a dry cleaner, Christa found time to join the glee club, drama club, German club, ceramics club, girls' basketball team, and student council, and to play a singing nun in a school production of *The Sound of Music*. An "average student" in her own estimation, she worked hard to make more As than Bs.

Classmates like Steve McAuliffe, the Clark Kent look-alike who became her boyfriend, spent senior year fielding college scholarship offers. Christa got none. A guidance counselor told her that a girl like her had four practical options: she could be a secretary, a nurse, a stewardess, or a teacher.

Christa couldn't type. She couldn't stand the sight of blood. The thought of flying made her queasy.

She told her boyfriend that she intended to be a schoolteacher. And one other thing: "If you asked me to marry you, I'd say yes."

He hoped she wasn't joking. "Will you marry me?"

"Yes," she said. "But we have to wait till we graduate college."

Steve was willing to wait. He accepted a scholarship from the Virginia Military Institute, six hundred miles away, and promised to stay faithful to her. Christa chose Framingham State College, a commuter school where tuition was only two hundred dollars a year. "Save your money for the boys," she told her parents, referring to her two younger brothers. "I'll live at home and get all the education I need."

At Framingham State, where she majored in education before switching to history, she never missed an 8:00 a.m. class taught by Dean of Women Carolla Haglund, "The History of Westward Movement." Campus gossips whispered that Haglund, who focused on the lives of the women and children history tended to forget, was a lesbian. Christa couldn't care less if Dean Haglund was a Martian; she was enthralled by Haglund's readings from the journals of women riding nineteenth-century wagon trains on the Santa Fe Trail, a thousand-mile trek from Missouri to New Mexico that took fifteen months. One pioneer woman wrote that she gave birth on the trail, "then I rode horseback and carried my baby on the saddle."

Between school activities, studying, and a part-time job waiting tables at Howard Johnson's, Christa kindled her long-distance romance by driving her Volkswagen Beetle through six states to visit Steve. It was a nine-hour drive in good traffic from Framingham to the VMI campus in Lexington, Virginia, but she said it was worth the trouble. When friends asked about their sleeping arrangements, she winked. On the way home, she often stopped in Washington, DC. Nineteen-year-old Christa Corrigan spent free afternoons sitting in the gallery during Supreme Court hearings or touring the National Air and Space Museum, looking up at Charles Lindbergh's single-seat airplane, the *Spirit of St. Louis*.

In her junior and senior years at Framingham State—"my radical years," she called them—she attended her first rock concert, a Jefferson Airplane show at Boston's Back Bay Theatre. She began wearing paisley dresses, white lipstick, and granny glasses. In 1969, she marched against the Vietnam War. She told her parents she was sorry if her activism made them uneasy but would not apologize for her beliefs. On graduation day, in 1970, she wore a black armband to protest the war.

Ed and Grace Corrigan's consolation came two months later. In a full-dress Catholic Mass and wedding at the Corrigans' home parish, Saint Jeremiah, three blocks from the house where Christa grew up, Steven James McAuliffe married his high-school sweetheart. Neither of them had gone steady with anyone else since they began dating at the age of fifteen. The bespectacled groom and his groomsmen wore white tuxes with black trim and black bow ties. The white-gowned bride had daisies in her hair. After their vows a guitarist strummed "A Time for Us," the love theme from the 1968 movie *Romeo and Juliet*.

——◆——

Christa took her husband's name. That was a choice she would second-guess for years. What kind of example was she setting, changing her name for no reason except that society expected it?

How would her husband feel about spending the rest of his life as Steve Corrigan?

At the same time, she loved her new name, the look and sound of it.

Christa McAuliffe

She had written the name a thousand times in schoolgirl journals and notebooks. Now it was hers, inscribed in her careful cursive loops on the ledger at the Publick House Historic Inn in Sturbridge, Massachusetts, where she and Steve spent their wedding night. The Publick House was "all antique," she wrote home to her mother. "There was lemon soap, and two apples on our bureau."

For the next three years, as Steve attended law school at George Washington University, Christa worked as a substitute teacher and waitress. She took night classes to earn her master's degree in secondary-school education at Bowie State, an inexpensive, historically Black college where she was one of the few white students. She and Steve had always said they'd return to New England when they got around to raising a family, but by the time their first child, Scott Corrigan McAuliffe, was born on September 11, 1976—"my Bicentennial baby," Christa called him—they had spent their first six years of married life in and around Washington, DC.

According to family lore, they were as happy as a sitcom couple until the following year, when Steve came through the door one evening with a surprise.

"Honey, I'm home! And guess what?"

His wife had taught a full day of classes, finished the housework, shopping, and laundry, and prepared their dinner.

"They want me at Justice!" Steve said. "A job in the Carter administration—isn't that great?"

Christa said, "You can live where you want, but Scott and I are going to live in New Hampshire."

So much for Justice. Steve looked for work in New Hampshire. He took a job in the state attorney general's office and they moved to Concord, the state capital, a town of thirty thousand built on the banks of the Merrimack River. "A Norman Rockwell kind of place," Christa called it. In 1979, she gave birth to a daughter, Caroline, who was named after two of Christa's heroines, her aunt Carrie and Caroline Kennedy, JFK's daughter. Caroline Corrigan McAuliffe and her brother, Scott, grew up in a brown-shingled three-story house their parents bought after Steve switched from the attorney general's office to a more lucrative private practice. Built in the 1920s, the house shivered when the wind blew.

Christa filled the place with heirlooms, including her grandmother's mahogany dining-room table, which must have weighed a ton, and tag-sale buys like a church pew she turned into a sofa. She relaxed with needlepoint or a copy of *Good Housekeeping*. Soon she and Steve hired a contractor to knock out part of the roof, put in a skylight and install a third-floor Jacuzzi where they could unwind after their workdays, looking up at the stars.

Christa went to work at Bow Memorial, a middle school near Concord. She became the most popular teacher there, a spirited lecturer who told her students there was more to history than old white men in paintings. "History's happening now. We're part of it," she said. She tacked *Time* and *People* magazine covers to the bulletin board: Ronald Reagan, Michael Jackson, Indiana Jones, the Mount Saint Helens volcano, the brand-new Rubik's Cube. She brought in a used-car salesman to tell teenagers how not to get cheated when they bought their first cars. She taught grammar and punctuation using the publication her students cared about most: the driver's manual.

"One thing I loved about her teaching was her ability to bring the world into her classroom," says her friend and fellow teacher Eileen O'Hara. "One day she walked into school carrying her books, papers, and a saucepan. A particular dish had come up in class and a few of her students had never heard of it. So she cooked a pot of it and brought it to school so her class could taste it."

As president of the Bow Memorial teachers' union, Christa McAuliffe announced that New Hampshire should be ashamed to rank forty-ninth out of the fifty states in education funding. The superintendent of schools didn't appreciate reading that quote in the *Concord Monitor*. "She thought it was a crime that teachers were paid so badly and women were second-class citizens," a friend recalls. Christa applied for the job of assistant principal at Bow Memorial, but the school board turned her down. The official reason: "No administrative experience." She told friends she was pretty sure she knew the real reason: they didn't want a woman on top.

——◆——

In 1983, she landed her dream job, teaching social studies at Concord High School. Everything about Concord High was "neat," high praise from her.

"She invited various professionals to speak to her students, people like the director of the New Hampshire ACLU," O'Hara recalls. "Through a program with the local bar association she got a volunteer 'Lawyer in the Classroom' to sit in on her classes and answer students' questions. Again she was connecting her classroom with the real world." Within a semester Christa had joined every faculty committee in sight and launched a frankly feminist social-studies course called "The American Woman." Fifteen girls and one intrepid boy signed up.

Like Carolla Haglund, Mrs. McAuliffe made textbook accounts "lively and even controversial," a former student says. Christa brought her guitar to class and sang sixties protest songs.

She had her students dress in period costumes and act out scenes starring Susan B. Anthony, Amelia Earhart, and Rosa Parks. Students voted on which women to study. One of their choices was Sally Ride, America's first female astronaut, who flew in the space shuttle *Challenger* that year.

Twenty years after she and Steve met at Marian High, the McAuliffes were living the life they had pictured as ambitious, sincere, slightly nerdy teenagers. Every weekday evening Christa tucked the children into bed, then brewed a cup of tea and took it to the living room. Sitting in front of the TV with the sound turned low, she sipped her tea and graded papers.

"Everybody loved Christa," says her college classmate Mary Liscombe, "but it's not like she was a saint." Mrs. McAuliffe fibbed on a lease application in Maryland, for instance, checking *No pets* after she and Steve adopted a cat they named Rizzo after the Dustin Hoffman character in *Midnight Cowboy*.

In Concord she organized a group of moms who'd pile into her VW van for raids on a Manchester grocery warehouse where they scored 40- and 50-percent discounts by posing as buyers for a supermarket chain. They bought fifty-pound sacks of flour and sugar, gallon jars of pickles, jugs of maple syrup, and spices at a discount, and divvied it up at a friend's house.

On August 28, 1984—a Tuesday—Christa picked up the *Concord Monitor* off the porch. The headline read "REAGAN WANTS TEACHER IN SPACE." A photo showed astronaut Judith Resnik, who had followed Sally Ride as the second female astronaut, climbing from the cockpit of a supersonic jet. According to the story, NASA was looking for a schoolteacher to fly on a space-shuttle mission. "Today, I'm directing NASA to begin a search," President Reagan had announced, "to choose as the first citizen passenger in the history of our space program one of America's finest—a teacher."

She didn't have time to read the whole story. She was pressed for time during her usual weekday-morning drill of whipping up four breakfasts, getting eight-year-old Scott out of bed, fed, dressed and ready for grade school, waking five-year-old Caroline and doing the same for her. After she'd showered and dressed she gathered up her daughter and the papers she'd graded the night before and drove to Concord High, where she dropped Caroline off at the school's student-run day care center. "Every morning we'd see her arrive—right before or just after the bell—with books and papers under one arm and Caroline in the other," another teacher remembered.

—◆—

Fifteen years after Neil Armstrong and Buzz Aldrin walked on the moon, NASA was looking for headlines. The glories of the Apollo program had ended when Apollo 17 astronauts Gene Cernan and Harrison Schmitt completed humankind's last moonwalk in 1972. Since then, with no extraterrestrial world within reach, the space program had shifted from missions of exploration to flights by four shuttles—"space trucks," some called them—to perform experiments and place satellites in orbit. The first shuttle missions had recaptured some of NASA's old glory, but within a year of Ride's news-making flight, shuttle launches had become so routine that the TV networks no longer carried them live.

Facing budget cuts from Congress, the agency considered sending a celebrity into space. But 1984 was an election year, and education was an election issue. The Reagan administration's budget cuts had led the National Education Association, which represented more than two million schoolteachers, to denounce Reagan as "America's Scrooge on education" and endorse his Democratic rival, Walter Mondale. With the election three months away, the president and his advisors saw a chance to promote the space program and win teachers' votes in one stroke.

"When that shuttle lifts off," Reagan announced, "America will be reminded of the crucial role that teachers and education play in the life of our nation. I can't think of a better lesson for our children and our country."

That fall, Christa and her friend Eileen O'Hara attended a National Council for Social Studies conference in Washington. "That's where she found the NASA booth advertising the Teacher in Space program. I wasn't surprised she picked up an application," O'Hara says. "She thought it would be a great way to influence students—not because it could make her famous, but because it was something unusual, something fun."

Returning with a sheaf of application forms, Christa passed them around to Concord High colleagues and sent one in herself. A week later, she found a shiny silver-and-blue package in the mailbox: NASA's official twelve-page application, designed to weed out anyone who might think of applying on a lark. The application called for lengthy answers to essay questions and multiple letters of recommendation. One newsman estimated that it would take "a serious contender" more than a hundred hours to complete.

Steve McAuliffe looked it over. Like millions of boys growing up in the sixties, he had dreamed of being an astronaut. He said, "Christa, you should go for it." But she didn't. Two months passed before he reminded her that the deadline was only two weeks away. "This is a don't-miss," he said. Not a can't-miss, considering the odds, but a long shot worth taking.

His wife disagreed. "They'll have scientists and PhDs," she said. Then, with time running out, she decided to go for it. She rounded up recommendations. She spent lunch breaks composing essays about her community involvement, communication skills, and philosophy of teaching, then tore up her essays and rewrote them on clean notebook paper in her tidy cursive. Until she noticed the fine print on the first page: *Please Note: Application form must be typed.*

She turned to her friend O'Hara, who had taken a job typing legal documents for Steve's firm. The three of them worked nights on Christa's Teacher in Space application. Christa would bring her handwritten pages to Steve's office after work and finish her essays with Eileen and Steve's help.

Why do you want to be the first U.S. private citizen in space?

"As a woman," she wrote, "I have been envious of those men who could participate in the space program and who were encouraged to excel in the areas of math and science. I felt that women had indeed been left outside of one of the most exciting careers available."

Steve read that over and said he doubted NASA was looking for some sort of women's libber. Christa pressed on.

"When Sally Ride and other women began to train as astronauts, I could look among my students and see ahead of them an ever-increasing list of opportunities," she wrote. "I cannot join the space program and restart my life as an astronaut, but this opportunity to connect my abilities as an educator with my interests in history and space is a unique opportunity to fulfill my early fantasies. I watched the Space Age being born and I would like to participate."

On the day of the deadline, running late as usual, she rushed her application to the post office. Hers was one of more than eleven thousand submitted by teachers all over the country.

"I know you'll think I'm silly," she announced to her students that day, "but I filled out the application to be the first teacher in space."

Later that month, after Concord High principal Charles Foley announced over the PA that Mrs. McAuliffe was "one of the seventy-nine teachers from New Hampshire to compete" in the program, a student asked for her autograph.

"Whatever for?" she asked.

The kid had his reasons. If she won, her signature would be a souvenir to keep forever. If not, it looked just like a hall pass.

2

———+———

EACH STATE SENT TWO CANDIDATES TO WASHINGTON FOR interviews with judges who would narrow the field to ten. New Hampshire's semifinalists were Bob Veilleux, a science teacher from Manchester, and Christa McAuliffe.

After Principal Foley announced the news on the PA at Concord High, the student who'd asked for her autograph crowed, "It's worth more already!"

Christa was packing for a night flight to Washington when the doorbell rang. "I had the kids down and was getting my clothes ready," she remembered. "A young girl in a highly agitated state stood on the porch. I knew her from school, although she had never been in one of my classes. She kept saying she was going to kill herself."

She brought the girl inside. Steve called that a bad idea. He said,

"You're not her counselor. There might be legal implications." Christa gave that a moment's thought. "So sue me," she said. She sat up with the girl for hours, listening, holding her hand. "By three o'clock we were both desperately in need of sleep," Christa recalled. "I put her to bed in Scott's room, in the lower half of his bunk bed."

In the morning Steve drove his bleary-eyed wife to the Manchester airport for her flight to DC, where NASA public affairs officers welcomed 114 Teacher in Space semifinalists—two from each state plus fourteen from US territories, international schools, and the Bureau of Indian Affairs—to the L'Enfant Plaza Hotel, where diplomats often stayed. Unfortunately for her, she was one of the last to arrive. There were no more rooms. The agency shuttled her to the Hyatt Regency on Capitol Hill, which struck Christa as plenty posh with its chandeliers, all-night room service, and TV screens that must have been two feet across.

The next morning the nominees gathered at the L'Enfant Plaza. According to Veilleux, "Christa was very excitable, very gee-whiz." They joined the others in the hotel's Grand Ballroom for a lecture by two astronauts: Joseph Allen, a veteran of two shuttle flights with a PhD in physics from Yale, and Judith Resnik, the second female astronaut. Resnik held a doctorate in electrical engineering from the University of Maryland. A veteran of ninety-six Earth orbits and 2.5 million miles in space on the shuttle *Discovery*, she was a black-maned beauty with an air of utter competence. Asked how it felt to ride a rocket into orbit, Resnik said, "It's like a fast, bumpy train ride. You feel shock waves, then your acceleration drops back. Six minutes later, you're in orbit."

Meeting real-life astronauts made Christa fret about her lack of training in science. Her credentials were no match for those of science teachers like Veilleux. And while she enjoyed hearing Resnik joke about being one of the test subjects who gave NASA's notorious spaceflight simulator its nickname, the "Vomit Comet," she couldn't imagine climbing aboard such a vehicle for a ride that made astronauts throw up. She doubted she had much chance of advancing to the final ten.

"It's a shame every one of you can't be taken," Resnik told the nominees. "I sure look forward to meeting the lucky person, and with any luck I'll be on that mission, too."

Another semifinalist, Massachusetts's Richard Methia, recalls feeling "out of my league" that day. Methia trudged from the hotel to the National Mall, "parked myself on a bench and took in the grandeur of the Capitol and the Washington Monument. Walking toward me was another teacher, carrying the same packet of materials emblazoned with the NASA logo." He thought she looked friendly. "I introduced myself. Christa sat down and we chatted. We both confessed to being overwhelmed by our competitors' achievements. She had a wonderful laugh. Pretty soon, like any two teachers, we were talking about our students."

NASA's blue-ribbon panel spent a week evaluating the candidates. The judges included Deke Slayton, the legendary chief of the Astronaut Office at the Lyndon B. Johnson Space Center in Houston; astronauts Gene Cernan and Harrison Schmitt, the last men on the moon; Dr. Robert Jarvik, inventor of the artificial heart; three university presidents; Washington Bullets basketball star Wes Unseld; and Pam Dawber, who played Mindy to Robin Williams's antic extraterrestrial on the TV comedy *Mork & Mindy*. When a reporter asked why on earth Dawber was on the panel, a NASA spokesman said, "Pam Dawber knows what it's like to become famous overnight." More than one of the teachers rolled their eyes at the idea of being graded by a sitcom actress, but many still lined up to get her autograph.

Veilleux didn't like his chances. "The judges didn't seem to want to hear from science teachers," he says. NASA had science up the wazoo. "They wanted a teacher who'd be good on the Johnny Carson show. Someone who could help make the public love space again."

The judges watched videotapes of the candidates. That was a boon to Christa, who came across on camera as her unaffected, upbeat self. As *Concord Monitor* editor Mike Pride put it, "The camera doesn't lie, and Christa didn't pretend."

Still, she didn't stick around for the announcement of the final ten. She flew home the night before, early for once. Steve met her at the Manchester airport and drove her home. They were in bed, asleep, when the phone rang at three in the morning. He picked up, listened for a second, handed her the phone.

"For you," he said. "It's NASA."

—◆—

The other nine Teacher in Space finalists included a Phi Beta Kappa Stanford grad, a language expert, a published poet who had founded a halfway house for troubled teens, a former air force pilot, and three accomplished mountain climbers. Six of the final ten were women, a fact that worried the men. One of the mountaineers, Kathleen Beres of Maryland, had crossed the Atlantic in a sailboat and was planning an expedition to Antarctica. In this crowd Christa felt like the suburban mom and intermediate skier she was. The *Washington Post* cited the others' achievements while describing her as "personable and spunky and enthusiastic."

President Reagan greeted the finalists in the White House on June 26, 1985. "Class will come to order," he said, and the teachers applauded. After congratulating them for being chosen from more than eleven thousand applicants, the president added, "I also want to tell you that your shuttle doesn't blast off for a while yet, so there's still time to back out!"

After that, NASA flew the final ten to Houston for a week of physical and mental tests at the Lyndon B. Johnson Space Center. With no four-star hotels nearby, the agency put them up at a Super 8 motel on NASA Road 1, a six-lane state highway through a landscape of strip malls, gas stations, barbecue joints, pizzerias, auto-repair shops, and minigolf courses. Not that the finalists had much time for recreation. Agency doctors poked and prodded them, took blood and urine samples, and put them through stress tests, lung-capacity tests, eye tests. "My gosh, they even know the height of my belly button," Christa marveled. After meeting a NASA proctologist she joked that she was "learning about parts of my body I never knew existed."

One finalist lost his chance to be the space teacher during an oxygen-deprivation exam. As the air in the chamber got thinner and thinner, Methia freaked out, fighting off NASA clinicians until they wrestled him down and pressed an oxygen mask over his nose. "I became extremely arrogant," he said afterward. "Had I been a pilot, I would have arrogantly crashed my plane."

Peggy Lathlaen of Texas felt herself panicking during the same test. She credited Christa with talking her through it. "Something about Christa was comforting," Lathlaen remembered. "Her eyes said peace and calm."

Up to then Mrs. McAuliffe's toughest recent competition had been the pro runners she'd followed in Boston's Bonne Bell Mini Marathon and the high scores she chased while chomping ghosts in her favorite video game, *Ms. Pac-Man*. Yet she sailed through the Teacher in Space preliminaries and aced her two-hour interrogation by Dr. Terrence McGuire, NASA's consulting psychiatrist.

"A lot of people don't see themselves as being okay," McGuire would tell *Monitor* reporter Robert Hohler. "Christa has a more objective view of who she is and what she's about. That doesn't mean she thinks she's perfect, that she isn't changing or doesn't want to change. But she has a good idea of who she is, and what she sees is pretty good. That's unusual today. I know this doesn't sound very scientific, but I think she's neat."

It was the claustrophobia experiment that worried her. Technicians called it the PRS test, short for "personal rescue sphere." The sphere, a nylon ball linked to an oxygen supply, was only a yard in diameter. Still in development for future shuttle missions, it was designed to allow crew members to survive an emergency that crippled a spacecraft but didn't destroy it. Christa dreaded climbing into this glorified, aerified beach ball. "I thought I'd start yelling and clawing to get out." She crept in and curled up while JSC workers zipped her inside. "It was dark and warm," she recalled. "I started fantasizing that I was lost in space." She relaxed. She hummed and sang songs to herself.

When the test ended she clambered out and asked if she could take her PRS home to Concord after this was all over. "Then, when things get crazy at home, I'll just set the timer and say, 'Okay, Mom's going into the sphere now!'"

Then came the last, most anticipated and dreaded test. Christa was pretty sure she would throw up on the Vomit Comet, which was not an earthbound simulator but a converted Boeing refueling plane known as a Stratotanker that flew roller-coaster parabolas over the Gulf of Mexico. During its steep dives, passengers experienced thirty seconds of weightlessness. They felt their bodies double in weight on the way up and then float against their harnesses as the plane flew almost straight down, leaving pens and sandwiches and lava-lamp globs of water hovering between the cabin's padded walls. Most riders vomited into barf bags. At a finalists' lunch with astronauts Resnik and Bob Crippen, a veteran of four shuttle missions and commander of three who was known as "Mr. Shuttle," Christa said she was sure she was going to throw up and ruin her chances.

Crippen told her not to worry. "Lots of astronauts get sick their first time."

Resnik corrected him. "Only the men get sick."

As the finalists prepared for their Vomit Comet ride, doctors handed out ScopeDex tablets. The pills were a blend of scopolamine, for nausea, and dexedrine. Christa and Beres, the Himalaya-climbing trans-Atlantic sailor, hesitated. Wasn't dexedrine an upper? If one of their students brought the dexedrine pills called "greenies" to school, they would have confiscated them. Still, neither teacher wanted to defy NASA doctors' orders. When in Houston . . .

Just as they swallowed the pills, Resnik pulled them aside. "Listen," she said, "that medicine they give you is too strong. You should bite it in half."

Too late! The teachers boarded the giant plane.

Over the next two hours, the Stratotanker climbed and dived twenty-seven times. During weightless half minutes the bolder finalists unclipped their harnesses and floated between the padded walls. Some did slow-motion somersaults. Some hung upside-down, tossing tennis balls and paper airplanes. They were delighted to discover that in zero G a paper airplane flies straight as an arrow. Christa joined the others long enough to form a gypsy-moth ring in the middle of the cabin, then retreated to her harness. She reached for a levitating Frisbee while her stomach turned. Gorge rising, she clamped her eyes shut. "It's almost over," Beres told her, squeezing her hand. Biting her lip, Christa made it back to earth without vomiting. The finalist from Vermont, former air force pilot Michael Metcalf, could not make the same claim.

After they landed, the astronauts treated the teachers to lunch at their favorite barbecue joint. A remodeled gas station, Pe-Te's Cajun BBQ House was thick with the pungent scents of cigarette smoke, wood smoke, pulled pork, sizzling ribs, molasses, and vinegar. Christa, still queasy, steeled herself and ordered gumbo. She must have looked a little greenish. Beres and some of the others kept trying to cheer her up, saying she was one of the front-runners. She thanked them and smiled, but she felt about as low as she was capable of feeling. One thing NASA didn't need, she figured, was a space traveler with a weak stomach.

—◆—

The finalists' handlers had one last treat in store for them: a field trip to the George C. Marshall Space Flight Center in Huntsville, Alabama.

Huntsville, a small city in the Tennessee River Valley, was the home of American rocketry. After World War II, a secret government program called Operation Paperclip had brought Wernher von Braun and more than 1,500 other German rocket scientists to Fort Bliss, Texas, to work for the US Army. Many of them, including von Braun, were former Nazis. They built America's nuclear arsenal, designing the missiles and guidance systems the army and air force aimed at the USSR and other targets around the world. In 1950, the army transferred the Germans to Huntsville, site of the Redstone Arsenal, a chemical-weapons plant. That helped turn the sleepy burg into a boomtown some Alabamans called "Hunsville." It made Madison County the proving ground for America's space program, where von Braun led the effort to launch men into orbit and beyond.

A quarter century later, Huntsville was home to the Marshall Space Center, named for the US Army general and secretary of state whose Marshall Plan had rebuilt postwar Germany. Along with the Johnson Space Center (JSC), in Houston, and the Kennedy Space Center (KSC) at Cape Canaveral, Marshall was one of three proud hubs of the American space program. It shared space with the U.S. Space and Rocket Center, site of

Space Camp, a theme park where the Teacher in Space finalists rode a space-shuttle simulator. Next came a bigger treat, a picnic with country-pop star John Denver. The singer-songwriter of hits including "Sunshine on My Shoulders," "Rocky Mountain High," and "Thank God I'm a Country Boy" was a self-described space nut who had applied for a spot on a future shuttle mission. Fresh off hosting that year's Grammy Awards, he told the finalists *he* was thrilled to meet *them*. "I'm going to write a song about the Teacher in Space," he said.

After the picnic, the finalists queued up for a seat in the park's spatial disorientation chair, which jerked and spun riders faster than the mechanical bull in *Urban Cowboy*. Few of the teachers wanted to go anywhere near the thing. "But we couldn't take chances. The stakes were too high," Beres remembered. "We were afraid that if we sneezed the wrong way we'd be in trouble."

Eight finalists rode the bucking chair without falling off. One staggered away to find a place to lie down. Only one got sick. Christa threw up into a Space Camp barf bag, then joined the others lining up for the last ride.

Lunar Odyssey was a domed enclosure similar to the Round-Up and other carnival attractions that whirl customers around a giant spindle—centrifugal force in action. Spinning faster and faster, like a runaway roulette wheel, it pressed riders backward with a force of three Gs. Lunar Odyssey also featured video screens showing liftoffs and space walks. "We are about to escape Earth's gravity," a voice intoned. Christa shut her eyes as the big wheel tilted and spun faster and the music grew louder. Then she heard a *thump*. "Stop it!" someone shouted. "Slow it down!"

"There was a shriek and repeated thuds," Methia recalls, "then dead quiet. All the lights flashed on."

As the music slowed and stopped, others chimed in. "Call 911!" "Call an ambulance!"

Their NASA handlers thought one of the teachers had been thrown off the ride. No—they were all accounted for, but Christa McAuliffe was calling for help. "Over here, over here!" They found her crouched beside a bleeding, still-breathing body.

—◆◆◆—

Twenty-year-old Todd Walker was a recent graduate of Huntsville's Virgil Grissom High School. He had a daredevil streak. Walker, who was working at the park, had decided to jump on the ride. But he put a foot wrong; a piece of Lunar Odyssey machinery caught his arm or leg or clothing and flung him through a partition. "We learned that the young staffer had been trying to impress us," says Methia. "He had untied his harness and was balancing on the high edge of the ride as it spun. He lost his balance." Now Walker lay in a heap near Christa's seat. Space Camp workers hurried to help. An ambulance sped the young man to a hospital, where he died.

Beres watched Christa follow their handlers to the exit, "crying uncontrollably." That evening, NASA flew the teachers back to Washington.

—◆◆◆—

Nineteen eighty-five was the first year of President Reagan's second term. With a little help from the Teacher in Space program, he had been reelected with 525 Electoral College votes to Walter Mondale's 13. Nineteen eighty-five was the year Microsoft released Windows 1.0, the year Michael Jordan was the NBA's Rookie of the Year, and the year Michael Jackson, Mick Jagger, David Bowie, Bruce Springsteen, Bob Dylan, Madonna, Queen, and U2 raised $127 million for famine relief in Africa at Live Aid, singing "We Are the World" and "Do They Know It's Christmas?"

The number-one movie was *Back to the Future*, released two weeks before the finalists reconvened at the White House to hear who would be named the Teacher in Space.

Reagan sent his regrets; he was recovering from colon surgery. That left Vice President George H. W. Bush to do the honors.

White-gloved waiters served lemonade in the Roosevelt Room. After introductory remarks by NASA administrator James Beggs, Bush nodded to the finalists standing at his side. Christa, her hair freshly permed, wore a skirt and blouse her sister Lisa had knitted for her, a string of pearls Steve had given her, and a golden silk jacket, a gift from her parents. "NASA, with the help of the heads of our state school systems," the vice president began, "has searched the nation for a teacher with the right stuff." Bush said there were "thousands, *thousands* of teachers with the right stuff." But only one would be chosen today. The finalists fidgeted while he studied his notes. At last he introduced "the winner, the teacher who will be going into space—Christa McAuliffe."

Bush turned to the finalists. "Where is she? Is that you?" Christa beamed, nodded. Blinking back tears, she accepted the others' congratulations. The vice president shook her hand. Returning to his notes, he announced, "Christa teaches at Concord High School in Concord, New Hampshire. She teaches high school social studies. She's been teaching twelve years. She plans to keep a journal of her experiences in space. She says, and here's the quote, '*Just as the pioneer travelers of the Conestoga wagon days kept personal journals, I, as a space traveler, would do the same.*' Well, I'm personally looking forward to reading that journal someday."

Bush called her forward. "Good luck, Christa," he said, handing her the Teacher in Space trophy, a bronze statuette of a student and teacher looking up at the stars.

During the applause that followed, Bush leaned toward her. "Do you want to say something?" he asked in a stage whisper. She surprised him by saying yes. When the applause ebbed, she stepped past him to the podium.

"It's not often that a teacher is at a loss for words," she said. "I've made nine wonderful friends over the last two weeks, and when that shuttle goes up, there might be one body . . ." Overcome for a second, she put a finger to her lips. "But there's ten souls I'm taking with me."

3

SHOPPERS STOPPED STRANGERS ON MAIN STREET IN CONCORD. "Did you hear? It's Christa!"

Steve McAuliffe paced in his law office, muttering, "Unbelievable . . . unbelievable." He phoned home to share the news with their kids. Nine-year-old Scott was thrilled. "She won?" The second grader saw a Disney opportunity. "Does this mean we get to go to Florida?"

"It does indeed," his dad said. "We're going to Florida!"

Caroline, who was going on six, hated the thought of losing her mom for months of space training in Houston, wherever that was. "It's not fair!" Steve won her over with talk of Disney World and her favorite meals, pizza and spaghetti. That evening she greeted reporters on their porch, saying, "I saw my mommy on television."

In a tavern on Route 9 in Massachusetts, a drinker seeing Christa on the bar TV said, "Hey, she used to be my babysitter!"

On Joseph Road, in Framingham, cars and TV trucks double-parked outside the house where the Teacher in Space grew up. Grace Corrigan answered the door between phone calls while husband Ed led a camera crew from Boston's Channel 7 to the living room, where he sat at the piano and sang one of their daughter's favorite songs:

Move over, sun, and give me some sky,
I got me some wings I'm eager to try,
I may be unknown but wait till I've flown,
You're gonna hear from me.

It was past midnight when Christa's return flight from Washington touched down at the Manchester airport, with its single terminal and blacktop runways. Several hundred well-wishers had stayed up late to welcome her home. A bagpiper played while a couple dressed as the "Pigs in Space" from TV's *The Muppet Show*, First Mate Miss Piggy and Captain Link Hogthrob, danced. Ed and Grace Corrigan hugged their daughter and handed her a dozen roses. Christa, feeling spent, had expected the usual half-hour drive home. Instead, a lights-and-sirens squad car led a caravan to Concord, with Steve and Christa's VW bus in the lead, followed by her parents in Grace's rusty Ford Mustang, followed by friends, relatives, and Christa fans happily honking their horns.

By the time they reached the interstate, Grace knew something was wrong. "We heard metal scraping and could see sparks flying," she recalled. "My muffler had come loose and was bouncing on the road. We would have loved to pull off, but fearing that those in back would follow us, we thought it best to stay in line."

THE BURNING BLUE ▪ 31

Rather than lead Christa's homecoming caravan to a dead end in Bear Brook State Park, the Corrigans made it to Concord, kicking up sparks all the way.

Every window in the McAuliffe house was lit up. Scott, awake way past his bedtime, asked his dad to send all the people away. Steve said, "Bunky, you can't just buy a ticket for a shuttle ride. This is a big deal!"

For two weeks their phone rang day and night. Movie producers wanted the rights to Christa's life story. "Can you imagine? What would a movie about me be?" she asked. "Boy meets girl, girl and boy fall in love, girl goes up in space?" There were offers for Christa McAuliffe merchandise. "People want me to help them sell balloons and T-shirts," she told an interviewer. "What I'd love would be to lend my name to an education fair or a Read-a-thon, something to get students interested in learning."

On August 6, 1985, the city of Concord held a Christa McAuliffe Day parade past New Hampshire's golden-domed statehouse. Steve took pictures while his wife rolled by in a gleaming black Mercedes convertible, flanked by Scott and Caroline, the three of them waving and giving thumbs-ups to tourists, friends, and neighbors. Downtown merchants hung signs in their windows: GODSPEED and GO CHRISTA GO! Soon she and Concord mayor David Coeyman climbed onto the back of a flatbed truck. The plump, mustached mayor checked his pocket watch, then announced, "We proclaim that the city of Concord wishes to extend our heartiest congratulations and highest commendation in seeking and achieving the honor and distinction of being the first teacher-astronaut!" He gave Christa a commemorative pewter plate and a cheerful embrace. "Whoa!" the mayor said. "I hugged an astronaut!"

Looking out at a crowd of friends and neighbors, reporters, curious shoppers, and pedestrians, Christa thanked everyone for coming out for her parade. Then she got to what she considered the point of her mission. "I am delighted to be a representative of the teaching profession," she said. "But it won't mean anything unless the adults out there support their schools. Unless the teachers out there truly believe in what they're doing. Unless you kids out there do the best you can and get the best education you can. So when I'm up in that shuttle and I'm not teaching at Concord High School, I want everybody working real hard to make education what it should be in this country!"

The mayor invited her to lead the town band in "Stars and Stripes Forever." As the *Boston Globe* reported, "McAuliffe, a self-described feminist and Kennedy Democrat in a conservative Republican state, returned to a homecoming parade down Concord's Main Street as adoring students and residents cheered her on. She handled instant fame with characteristic enthusiasm, grabbing the baton to lead the band's performance at a celebration in her honor."

According to *People* magazine, "You can hear America thinking, 'Christa, this could be the beginning of a beautiful friendship.'"

Not everyone agreed. The McAuliffes were taken aback when their life-insurance company, citing the risks of spaceflight, cancelled Christa's policy. Luckily for them, Lloyd's of London came to the rescue, scoring PR points with a free million-dollar policy. Then there was public criticism from John Glenn, the most iconic astronaut of all. Glenn, now a sixty-four-year-old senator from Ohio, dismissed the Teacher in Space program as a publicity stunt. A Democrat, he accused Reagan of sending novices like Christa McAuliffe on "cosmic carnival rides" for political purposes. Glenn said space was no place for "the butcher, the baker, and the candlestick maker."

Reagan's backers dismissed Glenn's attack as partisan politics. But Wally Schirra had no political axes to grind. The only astronaut to fly in the Mercury, Gemini, and Apollo programs,

Schirra was now chief of the Astronaut Office in Houston. He brought an engineer's view to the program. "The shuttle is not a passenger plane yet," Schirra said. "In fact, we're still learning to fly the thing."

—◆—

After NASA's public affairs officers (PAOs) booked Christa on the *Today* show, a hired car pulled into the McAuliffes' driveway to take her to the airport. In New York, a limousine met her flight and delivered her to NBC's Rockefeller Center studios, where a stylist powdered her face and gave her a hint of rouge. The next thing she knew she was on-camera with *Today* host Bryant Gumbel, who got straight to the point. "Why you?"

She smiled. "It's hard to say. I still can't really believe I'm going into that shuttle. It doesn't feel possible. Maybe when I'm on the launchpad it will."

"Maybe a little fright, too?"

"Not yet. Maybe when I'm strapped in and those rockets are going off underneath me, I will be!"

"Christa wasn't the best or the smartest teacher in America. What she had was a sort of genuineness that came across on TV," says Richard Methia. Soon, she was on all three network news programs. CNN named her one of its "Heroes of 1985." In interviews with publications including *Time, Newsweek, Ms.,* and *Ladies' Home Journal,* she said she wasn't sure which was more mind-boggling, the thought of flying into orbit or the idea that people saw her as a celebrity. *People* was preparing a Christa McAuliffe cover to coincide with the *Challenger* launch.

Back home, the *Concord Monitor* launched a Christa McAuliffe beat. "It was an amazing story and an amazing summer for Concord," recalls Mike Pride, the paper's editor at the time. "We threw ourselves into it." Pride assigned *Monitor* reporter Bob Hohler to be "Christa's shadow."

As Christa's own mother put it, "NASA wanted to rekindle the excitement that once surrounded the space program by choosing a teacher with the gift of gab." More than that, Grace Corrigan said, her daughter was "pretty but not too pretty, with wholesome attractiveness—the sort of good looks to which ordinary people can relate."

When a reporter asked Christa how she was managing a thousand new demands on her time, she said she was trying to take her own advice. On the first day of every term she told her students, "All I ask is that you be yourself and do the best you can."

—◆—

One of her biggest early challenges came on Johnny Carson's top-rated *Tonight* show. NBC flew her to Los Angeles. Arriving at the network's studios in the suburb Carson called "beautiful downtown Burbank," Mrs. McAuliffe from little old Concord found her name on a dressing-room door backstage.

It was "a little much" for a small-city schoolteacher, as she recalled. She tried to take it all in, from the sight of herself in the makeup-room mirror to the sound of Doc Severinsen's band playing her on.

Introducing her, Carson said, "I'm sure you have read or heard about my first guest tonight. She has a three-page article in this week's *People* magazine, and she was selected from a field of over eleven thousand applicants to be the first citizen in space." Next came a ten-second clip of her teary-eyed acceptance speech at the White House.

Carson said, "Please welcome Christa McAuliffe."

Walking from the wings into the spotlights, settling into the seat beside the host's desk, she heard applause that went on and on. How many nights had she spent grading papers with Johnny Carson's show on in the background? How strange to find yourself inside the TV with ten million people watching, including just about everybody in Concord.

Carson opened with a needle. "Christa, you didn't get emotional during that announcement," he joked. "You stayed cool and collected, like a real astronaut."

Was he trying to fluster her? He went on: "Who was it who once said—Deke Slayton, I think—'It's a strange feeling to realize every part of this space capsule was built by the lowest bidder!'" That line got a laugh. "What subject do you teach, by the way?"

"I teach an economics course," she said. "I teach a course I developed called 'The American Woman.'"

"And what are you going to be doing in space?"

"Teaching three lessons from the shuttle," she said.

Carson had a joke ready. "I remember a few teachers, when I was a kid, I would have *loved* to see go into outer space!"

Carson was known to make snap judgments. If he thought a guest was a blowhard or a phony, he would roll his eyes and go to commercial. But the longer they chatted, the more he perked up. "You're really excited to go, huh?" he asked.

"I can't wait."

Her time was up, but the king of late-night TV had a few parting words. He said, "NASA made a very good choice. I wish you well, and I think the whole country wishes you well." Severinsen's band played her off with a brassy rendition of the air force hymn: *Off we go into the wild blue yonder* . . .

—◆—

The McAuliffes had promised their friends the Bradleys that they would spend the last weekend of summer break at Hawks Nest Beach in Connecticut. Now her fellow teacher Carol Bradley said,

"I guess we can't go this year." Christa said, "Why not? We'll work it out." She was true to her word: on the last afternoon of the McAuliffes' and Bradleys' beach holiday, yet another limo pulled onto the sand to drive Christa to New York. Two hours later, she was in a ballroom at the Grand Hyatt Hotel on Forty-Second Street, mingling with ambassadors and TV stars at the 1985 Emmy Awards for news and documentary programs, presenting an Emmy to ABC anchorman Ted Koppel.

She had a week to spend at home before reporting to Houston. She cooked, cleaned every corner of the house, and returned to Concord High for a convocation to kick off the 1985–86 school year. Welcoming the class of '86, the world's most famous schoolteacher said she was going to be sorry to miss the seniors' last year. "But don't worry. I'll be back here to speak at your graduation." They whistled and cheered and unfurled a banner: MRS. MCAULIFFE . . . HAVE A BLAST!

The worst part would be leaving five-year-old Caroline behind. According to Hohler, the *Monitor* reporter serving as Christa's shadow, "Caroline had cried each time her mom had gone away." Hohler dropped by the house for spaghetti dinners in the McAuliffes' dining room with its chandelier and floral wallpaper, its Persian rug and cabinets holding Christa's heirloom china. Her bookshelves held well-thumbed copies of *The Feminine Mystique* and *The Women's Room.* As modern parents, she and Steve were determined to explain each step of her journey to the kids. "Christa wanted to prepare her children for the fire and thunder of a shuttle launch," Hohler reported. She had heard of how astronaut Anna Fisher's two-year-old daughter shrieked in terror when the shuttle *Discovery* took off, thinking her mother was burning up. "We're all in this together," she and Steve told their children. Someday, they said, the four of them would look back together on Mom's great adventure.

Scott was on board from the start, but Caroline still had her doubts. After one of Christa's overnight trips, Steve drove the kids to the Manchester airport to meet her return flight.

A nice lady asked the little girl who she was waiting for. "Mommy," Caroline said.

"Is this the first time she's been away?"

"No. She does this a lot. She comes and goes and comes and goes. She just wants to go into space or something."

———◆———

The Lyndon B. Johnson Space Center occupies two and a half square miles southeast of Houston, just inland from Galveston Bay. Before NASA selected the site in 1961, this was cattle-ranching land. By the 1980s, it was the busy workplace of more than twelve thousand engineers, executives, technicians, astronauts, and other space-program employees. When Christa McAuliffe reported for duty on September 8, 1985, she found a grid of concrete-and-glass office buildings, man-made ponds, and Bermuda grass lawns on streets named Beta Link, Gamma Link, and Saturn Lane, all protected by ten miles of chain-link fence topped with barbed wire. She and Barbara Morgan, the second-grade teacher from Idaho who would be Christa's backup as Teacher in Space, showed their driver's licenses to a guard, who raised the barricade at the main gate.

Morgan, the Phi Beta Kappa graduate of Stanford, would never forget how she felt the day they checked into their spartan apartments near the Space Center: Thrilled. Nervous. Jumpy. She was hanging up her clothes, wondering what she had gotten herself into, when her doorbell rang.

It was Christa. She had baked chocolate-chip cookies for the two of them.

In the morning they reported to Building 4-S, where TV crews filmed the Space Teacher's arrival. For two months Christa had weathered the media storm gathering around her mission. Now she was on NASA's turf, entering a warren of offices that had belonged almost exclusively to men since the dawn of the space program in the 1950s. Before 1978, when the agency admitted six women, three African Americans, and an Asian American to the roster, there had been seventy-three astronauts in NASA history; all seventy-three were white men. The *Washington Post* had hailed the arrival of "women and other minorities" despite the fact that women made up more than half the country's population. And though half a dozen of them were women, the 1978 class of "AsCans"—NASA shorthand for "astronaut candidates," pronounced "AssCan"—was quickly dubbed the "Thirty-Five New Guys." Seven years later, some veteran astronauts still snickered at what one of them called the "equal-opportunity pioneers" in their midst.

Christa wrote home:

Dear Mom & Dad,
Made it to Houston—2 planes were late and we had a hard time finding the rental car, but Barbara and I are having a good time. . . . Thanks for the throws for Scott's bed—he loves them and so does Rizzo! I'll write again.

She fretted about meeting the rest of the crew until she shook hands with Commander Richard Scobee, the veteran astronaut who would command her flight. "He seemed like the astronauts of her youth," Bob Hohler noted, "wholesome and handsome, tall, blue-eyed and ruggedly built, with a square jaw and an air of self-confidence." Dick Scobee introduced Christa to his wife, June, a lively University of Houston professor. "Here's the one with the important job," he said. "June's a schoolteacher."

June Scobee embraced the newcomer. "I'm terribly jealous! I wish I could fly, too, but I came along too late," she said.

Christa wrote home about the commander's wife: *Mom, you will just love her. She understands where I'm coming from. She's a teacher—I can talk to her!*

On her second day in Houston, Christa picked up her flight gear: a pair of black boots for the launch; leather-soled moccasins for zero G; rubber pants to keep her blood circulating during reentry; a sky-blue flight suit with her name over the heart; a red, white, and blue space helmet. She taste-tested food and beverage options for her week in orbit, giving a powdered strawberry drink a score of five on a scale of one to ten. Vacuum-packed rice pilaf got a seven, broccoli and cheese a nine. She admitted she was grading on a curve: "Compared to cafeteria food, it's terrific."

Their mission was officially named STS-51L. STS stood for "Shuttle Transportation System." The rest was alphanumeric soup that dated back to a superstition of Deke Slayton's. Slayton, the Mercury astronauts' crew-cut guru and longtime chief of the Astronaut Office at JSC, was a triskaidekaphobe; the number thirteen gave him the creeps. The first shuttle flights had been dubbed STS-1 and STS-2, but as the numbers mounted the agency came up with a new system. Beginning in 1984, missions were designated STS plus a number in which the first digit stood for the final digit of the mission's fiscal year, the second stood for the launch site (1 for KSC, 2 for Vandenberg Air Force Base), and a letter represented the order in which a shuttle would launch during the mission's fiscal year.

By the old method, Christa's mission would have been STS-25. Instead it became STS-51L.

Her main task on the shuttle would be conducting a pair of science lessons that PBS would beam to classrooms all over the

country. Her NASA handlers dubbed them the "Ultimate Field Trip." She would also keep a journal of her adventure, a sequel to the journals of the pioneer women she admired. Her contributions to the mission would be living proof of one of Christa's classroom slogans: "Ordinary people make history."

—◆◆◆—

Eileen O'Hara, her replacement at Concord High, thought Christa's story would make a good topic for her class on the American woman. Others saw her as a celebrity. "My gosh, I haven't done anything. Ask me after I've flown," Christa said. Later, O'Hara would find several journal pages in her friend's desk. Before leaving for Houston, the Teacher in Space had taken time to jot down a few notes—not about schoolwork but about Scott and Caroline. "*They are so different, my children,*" she wrote, describing her son as "*delicate . . . so sensitive that he turned off Sesame Street during an episode in which a cartoon mouse was eaten by a cartoon cat.*" Caroline, by contrast, got clingy when her mom left home but was otherwise "*Miss Independence. A gust of energy, ready for anything. It is a constant amazement to my husband and me that the same genetic structure could produce two such unique individuals.*"

Her hubby, as magazines called him, took on his new duties with pride and puzzlement. "*Steve seems to be holding down the fort,*" she wrote to her parents. "*He's been trying out the microwave—I even had to listen to it beep the other night (Scott insisted). . . .*" Microwave ovens were still new. Christa had bought one before jetting off to Houston but hadn't had time to help Steve learn to cook with it. He told her not to worry—the owner's manual couldn't be too much harder than the bar exam.

"His office is small and spare," the *Washington Post* reported, without noting that President Franklin Pierce, another New Hampshire lawyer, had once occupied the same space. "Before the *Challenger*, it was McAuliffe and not his wife who was the better-known half of the pair—an up-and-coming lawyer at one of the most prominent and prosperous firms in the state. A Georgetown University Law School graduate, McAuliffe came to New Hampshire with his wife in 1978 and went on to become a partner in a private firm whose well-appointed offices sit within sight of the gold-domed state capitol." Now, to Steve's surprise and frequent amusement, he was Mr. Christa, walking B-roll crews through their house while she chatted with Johnny Carson on national TV.

Her parents drove up to Concord to help with laundry and errands and answer the phone. Neighbors dropped by with casseroles and legs of lamb. Steve's life changed almost as radically as his wife's, and he welcomed the change. "Before all this, I was one of those fathers who always came home after the kids were in bed," he said. Now he helped Scott and Caroline dress for school on weekday mornings. He whipped up their breakfasts and packed their lunch boxes. After school he joshed with them when dinner went up in smoke and he ordered pizza instead. They shared the phone during Christa's nightly call, and he tucked them into bed afterward. "Whatever the price of readjustments—taking on things I probably should have been doing anyway—they pale in comparison to her opportunity," he said.

Challenger was scheduled for launch in January 1986, only four months away. Christa had twelve-inch-thick training manuals to study. She had mental tests to finish, treadmill tests, vision tests. She told her PAOs that she wanted to spend more time training and less on public appearances. Still, there were opportunities she wouldn't dream of passing up. One was a state dinner at the White House, with Steve as her guest.

Looking around the State Dining Room in the West Wing, she recognized movie stars Raquel Welch, Sylvester Stallone, and Michael J. Fox, and singer Natalie Cole. She met the president of Coca-Cola. "But of all people," she marveled, "the president invited a high-school teacher to sit beside him!" Over jellied sole and sunflower salad, President Reagan asked Christa how her training was going. They swapped stories about his movies and the televised lessons she would teach from the shuttle. Then he leaned toward her and said, "Watch this." With that, the president stood up, and in a matter of seconds everyone in the room, following his lead, stood up. Reagan winked at Christa and said, "How about that?"

That night, she and Steve felt far more than five hundred miles from parent-teacher potlucks at Concord High. "Steve spent a few extra minutes with Raquel Welch," Christa told a friend, "and I could see why!"

A reporter for *Space News Roundup*, a NASA newsletter, asked Christa about her sudden fame. Did she expect to get rich? "Oh, help!" she said. "You are talking to a teacher. I didn't choose my career so I could get monetary rewards. I never would have gone into teaching if that's what I wanted!"

At her next press conference, reporters hurried past the rest of the crew to get to Christa. Judith Resnik told her, "You know, you don't have to put up with all this."

Christa said, "You don't have to in your job, but I do in mine."

The agency assigned US Army colonel Bobby Mayfield to help the Teacher in Space prepare for her lessons from orbit. Mayfield had his doubts about the assignment. He tried to give Christa the benefit of the doubt—anybody would have been dazzled by the spotlight that struck her the moment she was named America's Teacher in Space. In Reagan's words, Christa McAuliffe was the best advertisement NASA ever had. But in Mayfield's view she

was a high-school social-studies teacher who had trouble understanding the science in her fifteen-minute scripts. He knew she
was exhausted—so why not push back, say no for once? Mayfield
saw Christa's saying yes to every interview as a lack of commitment to the mission. He resented it. He resented her.

4

TRAINING STARTED AT EIGHT SHARP EACH WEEKDAY. CHRISTA and Barbara Morgan jogged on treadmills for technicians who checked their heart rates and oxygen uptake. Christa, who was no sprinter, said her goal was "to make sure I don't fall off the treadmill." Other tests measured the teachers' vision, hearing, and reaction time. She failed an eye test and spent a long day worrying that it would disqualify her. Only one astronaut had worn contact lenses in space, and that was as an experiment. In the end, the agency made an exception for her.

She and Morgan spent hours on the space center's computer-aided instructional trainer (CAIT) and crew software trainer (CST), familiarizing themselves with the shuttle's workings. They studied the mission's four-thousand-page flight data file. While career astronauts occupied the Astronaut Office on the top

floor of Building 4 at JSC, the schoolteachers occupied tempo-
rary quarters in another building. Their office featured the same
government-issue swivel chairs, dented filing cabinets, and fluo-
rescent lights, but lacked the personal touches the Astronaut
Office was known for—souvenir footballs; Nerf basketball hoops;
Playboy pinups; bumper stickers and pennants promoting Army,
Navy, Purdue, MIT, and flyers' other alma maters; as well as a bulletin board known for the irreverent comments posted on it. With
female astronauts now flying, one bulletin-board note defined the
word "cockpit" as "a woman." Another joker appended a note to
a magazine story on "sex in space," claiming outer-space inter-
course was impossible because Newton's Third Law of Motion—
for every action there is an equal and opposite reaction—would
send lovers bouncing off each other in opposite directions. Beside
that, someone scrawled *No! This is why God gave us arms and
legs.* Below that was a sign-up list for sex-in-space volunteers to
which someone, thought to be one of the six female astronauts,
added yet another note: *Grow up.*

McAuliffe and Morgan's Teacher in Space office was bare
except for two props the agency had chosen for them, a map of
the world and a blackboard. Someone had written a greeting on
the blackboard: *HAVING FUN YET?*

Christa personalized her desk with framed photos of Steve
and the kids. She tacked up a drawing Caroline had made of her
mommy in a spacesuit. Looking around at the office decor, which
one visitor described as "early Holiday Inn," she joked that she
knew why the agency had picked the two of them: NASA was
paying their teacher salaries for the year, and their home states,
New Hampshire and Idaho, were at the bottom of the list in that
department. "They're saving money on us!" She and Morgan
spent office hours studying, comparing notes about their training
sessions, and memorizing some of NASA's many abbreviations:

A/L for "air lock," C/L for "checklist," D&C for "display and con-trols." A spacewalk was an EVA, short for "extravehicular activ-ity." When nature called in orbit, shuttle crews floated to a $1.2 million vacuum-powered WCS, or "waste collection system," or toilet. When Morgan discovered that the agency's official list of more than 120 acronyms carried a footnote directing readers to "a more complete list in NASA Reference Publication 1059," she coined a new name for NASA: the National Agency for the Sug-gestion of Acronyms.

Most of the career astronauts lived in subdivisions near JSC. The agency put McAuliffe and Morgan up at the Peachtree Lane Apartments, a cluster of tan stucco buildings three blocks from the space center. Their furnished apartments overlooked a Burger King, a Putt-Putt golf course, and a pair of gas stations. Once a week, an exterminator came by to fumigate, which barely dis-couraged the inch-long cockroaches the locals called "palmetto bugs." Some of the bugs took to the air when the poison stirred them up, flying from room to room.

At first Christa was pleased to see picnic tables and a swim-ming pool below her balcony, but Houston's heat, even in Septem-ber and October, got so oppressive she stayed inside. She missed home. She missed leaf-peeping at this time of year. She set the thermostat at 65 degrees and wore jeans and sweaters around the apartment, the way she used to do at home. She ended every day the same way, sipping a cup of tea with the TV on in the back-ground, studying training manuals and writing letters home. The upside of Houston's heat and humidity, she wrote to her mother, was that she was saving on perms: "Your hair frizzes when you walk out the door!"

Sometimes she joined the career astronauts and NASA work-ers at the Outpost Tavern, a barn behind a gravel parking lot.

Inside, the walls were covered with photos of shuttle launches and shuttle crews. A bulky flight suit hung from one wall, as if its wearer had jumped out and joined the crowd at the bar, where Robert Lee "Hoot" Gibson could often be found entertaining fellow astronauts, their spouses, lovers, and friends, as well as the bartender, who served up Gibson's favorite drink, a flaming hooker.

Gibson was a blond navy flyer with a bristly mustache. He had served as *Challenger*'s second in command a year earlier and was scheduled to command the next flight of the shuttle *Columbia* in December 1985. One of the more colorful members of the Thirty-Five New Guys chosen by NASA in 1978, he carried on the macho tradition of spacemen that dated back to Alan Shepard, the Mercury Seven's original horndog. When Gibson passed a good-looking woman he would make a noise like a pig and say, "Golly, I'd like to snort her flanks." Marrying fellow astronaut Rhea Seddon in 1981 had calmed him down some, but Hoot was still the Outpost's leading fire-eater. His favored flaming hooker was a brandy snifter filled to the brim with high-octane liquor and served on fire. A rum like Bacardi 151 was more than 70 percent alcohol—less flammable than rocket fuel but not so far short of gasoline. The idea was to finish the drink before it burned part or all of the drinker's face. Those who failed got razzed for the Band-Aids and singed mustaches they sported the next day.

"Be fearless!" Gibson commanded. His advantage was knowing the physics and chemistry of the drink's path from snifter to gullet. "You have to toss the whole glass." Fire requires oxygen, "and there's not enough oxygen in your mouth to feed the flame, so it goes out—*if* you do it fast enough."

"Needless to say," another astronaut recalled, "it helped to be at a bulletproof level of intoxication before attempting this trick."

While others cheered him on, Gibson downed his flaming

hooker, banged the still-smoking snifter to the bar, and turned to the crowd:

"Who wants the next one?"

Christa and Morgan stuck to beers and margaritas. As they got to know Greg Jarvis, another non-astronaut assigned to the *Challenger* crew, they invited him along on trips to the Outpost and the supermarket. Like Christa, the Hughes Aircraft engineer was not a full-fledged astronaut but a "payload specialist," a title created for the civilians NASA had added to shuttle crews starting in 1983. Jarvis's ride on *Challenger* would be a perk for Hughes Aircraft in return for its vast investment in the space program. Like Christa and Morgan, he knew how it felt to be an outsider among the career astronauts. Some nights he joined them for a game of Trivial Pursuit.

On one of Christa's press junkets, a reporter asked if she was no more than a trivia answer herself: "Are you a publicity stunt?"

She thought about that. Critics of the Teacher in Space program suggested that the $650,000 her training was costing taxpayers might be better spent on teachers' salaries. One reporter dismissed Christa McAuliffe as "a pawn of the Reagan administration, used to gain favor with teachers and by NASA to gain favor for the space program."

She saw her mission differently. "You know," she said, "I think this is the best public relations gift to the schools that NASA could ever make. It's a bargain when you think of the students and teachers we'll reach." As for teachers' salaries, "I make a little over twenty thousand dollars. I think that after twelve years of teaching I should be making more. Are teachers' salaries competitive? No." A friend of hers had recently been named New Hampshire Teacher of the Year, "and she makes twenty-three thousand dollars after forty years on the job." A school district "right here in Houston" was hiring housewives and out-of-work businessmen to fill in because there weren't enough young people going into teaching. "I'm representing my profession," she said. "I hope what I'm doing gets people to take a look at the tremendous contribution teachers make to our country."

In addition to poring over the training manuals and checklists NASA prepared for her and editing a study guide the agency would send to tens of thousands of teachers, Christa spent hours writing recommendations for Concord High students' college applications. She never lied but sometimes pushed the envelope by overpraising a B student, knowing how much a letter from the Teacher in Space might impress an admissions officer. And she never missed her nightly phone call to Steve and the kids. Each evening, after talking with Steve about everything from grocery lists to their guest list for the shuttle launch in January to the parent-teacher meetings he now attended, she spoke to both children.

Scott hoped she would take his favorite plush toy, a frog named Fleegle, on the space shuttle. "I'll find out," she promised. She thought the agency probably had a policy, if not an acronym, for amphibians in orbit.

One night, Caroline asked, "Mommy, are you in space yet?"

"Not yet, honey. I still have some homework to do."

—◆—

Her trainers put her through most of the same paces professional astronaut candidates endured. On the appointed day early in their training, she and Morgan reported to a conference room at JSC to hear the Apollo 1 tape.

"Roll the audio, please," one of their handlers said.

First, they heard prelaunch chatter recorded on January 27, 1967. After the usual back-and-forth radio talk peppered with static, astronaut Virgil "Gus" Grissom griped about the comm link between the Apollo 1 capsule and several control rooms at Kennedy Space Center. "How are we gonna get to the moon if we can't talk between two or three buildings?"

Almost exactly a minute later, a spark ignited the pure-oxygen atmosphere inside the capsule. Pilot Ed White cried, "Fire in the cockpit!" The rest was brief but terrible.

"We have a bad fire!"

"Get us out!"

"We're burning up!"

There were screams, then radio silence. Despite their burns, Grissom, White, and Roger Chaffee died of smoke inhalation.

At first, NASA claimed the Apollo 1 astronauts had died instantly; only later did the agency tell the full story and use the accident as a cautionary tale during training. By 1985, no AsCan could say he or she wasn't warned how dangerous spaceflight could be.

—◆—

Colonel Bobby Mayfield was a middle-school teacher from Round Rock, Texas. His title at the Johnson Space Center was "Aerospace Education Specialist." At first, he welcomed the chance to help Christa prepare for the televised lessons PBS would beam to classrooms all over the country. But it didn't take long for him to discover that the two of them were oil and water, Lone Star beer and maple syrup. He was punctual; she was always running late. He was focused on the lessons and in-flight experiments she needed to learn; she kept being pulled in two or three directions at once, missing study sessions to talk with reporters and smile for the cameras. Even his accent grated on her ears, which she persisted in calling his "ee-ahs." According to him, "She spoke a different dialect."

As Christa struggled, Mayfield simplified the scripts NASA had prepared for her. Still she flubbed her lines. "This would be a lot easier if she knew science," he said. He reminded her that her duties were child's play compared to the astronauts' tasks. Nobody was asking her to fly the shuttle. She didn't have to know electrical engineering like mission specialist Resnik or laser physics like mission specialist McNair. Her job was to memorize two fifteen-minute scripts and prepare half a dozen lessons on the effects of weightlessness on plant growth, magnetism, and effervescence that would be videotaped and distributed to thousands of schools after *Challenger* returned to Earth. The lesson on effervescence might even be fun: it showed the odd behavior of bubbles in water at zero G. "Alka-Seltzer in space," Christa called it. Beyond that limited syllabus, her role during the shuttle's week in orbit would consist mostly of keeping a journal and trying not to bump into the real astronauts.

"This is Christa McAuliffe, live from the *Challenger*," she recited in one rehearsal. "I'm, um, going to introduce you to two very important members of the crew. The first is Commander Scobee, sitting to my left, and the second one is Michael Smith. Pilot Smith is going to tell you a little bit about SPOC, which is the computer that is used onboard." The computer's name stood for "shuttle portable onboard computer," but everyone thought of it as "Spock," a nod to the emotionless science officer on *Star Trek*. Spock and SPOC were both immune to the anguish Christa felt during her halting attempts to explain magnetism and chromatography.

"I've been lost sometimes," she admitted. "I've been told and reminded, told and reminded."

Almost as difficult for a positive thinker like her was fretting that the rest of the crew thought she was "thumbing a ride" on the space shuttle. After all, there were only so many seats on each billion-dollar shuttle mission. Hers could have gone to a professional—a career astronaut who could perform more challenging tasks than narrating a few elementary experiments and learning how to anchor a tube of toothpaste to a strip of Velcro to keep it from floating away. Many of the "real" astronauts had put in seven years of training while waiting for their hard-earned chance to turn the A-shaped pins NASA awarded them from silver, which identified qualified members of the astronaut corps, to gold, which marked the chosen few who had seen Earth from space.

Before Christa, twenty-two civilians had flown as payload specialists on shuttle missions. They were members of the Space Flight Participant Program (SFPP), which served two purposes that meant a great deal to the agency in an age of budget cuts: politics and publicity. In the spring of 1985, the crew of the shuttle *Discovery* had pitched in to help Senator Jake Garn, a Republican from Utah, play astronaut during seven days in orbit. Garn's qualifications? He was a former navy pilot and member of the state's Air National Guard. More important, he represented the state where Morton Thiokol, a NASA contractor that made the shuttles' rocket boosters, was based. Still more important was that Garn chaired the Senate subcommittee that controlled NASA's budget. A second spacefaring politician, US Representative Bill Nelson, a Florida Democrat whose district included Cape Canaveral, was chairman of the House subcommittee on science and space. Nelson was scheduled to fly on *Columbia* in January 1986, two weeks before *Challenger*'s Teacher in Space mission. A devout Christian, Representative Nelson claimed that the Lord had a plan for him to fly into space. He announced that he would be "looking for angels" in outer space. He wasn't joking, but his crewmates were when they worked up a trick to play on him during the flight. They were planning to set off alarms and say, "Oh no, an angel got sucked into our engines. Don't look—there's feathers everywhere!"

Sultan bin Salman Al Saud, a member of the Saudi royal

family, flew on *Discovery* in June 1985 as a reward for his country's multimillion-dollar deployment of a communications satellite, Arabsat-1B. Another payload specialist, Taylor Wang, was chosen for a *Challenger* crew to represent a key NASA partner, his employer, the Jet Propulsion Laboratory at Caltech. His specialty was manipulating liquids without containers at zero G, using exquisitely controlled sound waves instead of boxes or bottles. When Wang's experiment failed in orbit, he panicked. He shouted over the radio to Mission Control, "If you guys don't give me a chance to repair my instruments, I'm not going back!" According to his crewmate Mike Mullane, the episode "severely depressed him and he surrendered to episodes of crying." Wang later admitted that he considered killing himself, "but everything on the shuttle is designed for safety. The knife onboard can't even cut the bread. You could put your head in the oven, but it's really just a food warmer. If you tried to hang yourself with no gravity, you'd just dangle there like an idiot!" Fortunately for Wang and the program's reputation, Mission Control gave him time to repair his equipment and he gave up thoughts of suicide.

Strange things happened in orbit. Hoot Gibson marveled at the sight of "a razor blade floating in midair." *Discovery* commander Hank Hartsfield recalled a payload specialist who became "obsessed with the hatch. He said, 'You mean all I've got to do is turn that handle, the hatch opens and all the air goes out?'" Hartsfield didn't like the sound of that. "It was kind of scary, so we began to lock the hatch."

Still, NASA decided the Space Flight Participant Program was worth the risks. The first few shuttle missions had made headlines. Sally Ride's flight in 1983, the first by a female astronaut, made worldwide news and drew thousands of spectators to Cape Canaveral, but two years later, after twenty shuttle flights, shuttle launches seemed routine. The three TV networks had stopped carrying live coverage of launches. That's where Christa came in. Her upcoming ride on *Challenger* put NASA back in the spotlight.

To Judy Resnik, giving shuttle seats to politicians, princes, space-industry people like Greg Jarvis, and media darlings like Christa McAuliffe was the very definition of selling out. According to Hoot Gibson, "the mission specialists hated the payload specialists, because they saw them—and in a way they were— taking a seat that our mission specialists could have had or should have had." It was Resnik who said aloud what other astronauts were thinking: "What are we going to *do* with these people?" The answer: Babysit them. Coach them. Make them feel part of the team and, if necessary, keep them away from the hatch.

It was all Resnik could do not to roll her eyes at the reporters, photographers, and NASA PAOs chasing Christa around. A few of them even asked for Christa's autograph. To Resnik, the Teacher in Space was a walking, talking publicity stunt. At least at first. As weeks passed and she saw payload specialist McAuliffe keep her wits about her while coping with a torrent of media attention, she began to think differently. It couldn't be easy to be a spokesperson for the space program and America's three million teachers while bunking in a crummy apartment two thousand miles from her husband and children. The more Resnik saw of Christa, the better she liked her. And so, in the early fall of 1985, she took an engineer's approach to the problem in front of her. Seeing Christa struggle to master the college-level science and math she needed to learn, Resnik added another line to her own checklist for the twenty-fifth shuttle flight: unofficial science tutor to the Teacher in Space.

They met over coffee and between training exercises. "It's not as hard as they make it sound," Resnik said. Magnetism is caused by electrons spinning in the same direction. Space shuttles fly because two of the most explosive substances in the world, liquid hydrogen and liquid oxygen, combine to drive them into orbit, and the shuttle engines' exhaust—countless molecules made of two hydrogen atoms and one oxygen atom—is water. "That's why the shuttle's exhaust looks like steam," she said. "It *is* steam." For her part, Christa admired the veteran astronaut, with her raunchy humor and a mind that seemed to solve problems faster than Spock, the onboard computer. They could hardly have been more different: the happily married New England Catholic who thought "hell" was a cuss word and the first Jewish astronaut, a raven-haired divorcée who cursed a blue streak when something annoyed or delighted her. Christa was naturally gregarious and enjoyed talking to reporters. Resnik loathed the press because nine out of ten reporters focused on her looks and gender and had no interest in her work.

—▪◆▪—

Christa praised her new tutor in a letter on NASA letterhead:

> Dear Mom & Dad,
> Houston is hot and steamy. The crew is great. Judy Resnik (J.R.) especially has been very helpful. She has flown before.

Her sessions with Mayfield started going better. Years later, reflecting on how close he came to giving up on her, Mayfield said, "Christa McAuliffe turned out to be an A-plus student. My feelings turned from resentment to total admiration."

5

CHRISTA WAS SUPPOSED TO KEEP A JOURNAL WHILE SHE trained and traveled in space. According to her application for the Teacher in Space program, her journal would be a space-age version of the pioneer accounts that spurred her interest in American history at Framingham State. But she never quite got around to writing it. She gave interviews to local and national reporters. She wrote letters, postcards, and holiday cards to Steve, Scott, Caroline, and her parents, friends, and fans, all in her careful cursive. To Carolla Haglund, her mentor at Framingham State, she wrote of how she hated being sixteen hundred miles from her home and family, "but we realized that it was a tradeoff for a chance of a lifetime."

She often wrote to Eileen O'Hara, who took over her American Woman class that semester:

Dear Eileen,
I'm glad that you are having a good time with the social studies
department—they are all such nice people—a little strange at
times but aren't we all?

Writing in a spiral notebook she'd brought from Concord for
the purpose, she tried to start her "space journal." She would write
a few paragraphs, read them over, cross them out. She was always
more of a talker than a writer. Chatting with Bryant Gumbel or
Johnny Carson on TV was easier than writing lines that thousands
or millions of people might read.

Then there was the matter of who would publish her journal.
Steve, studying the contract she had signed with NASA, told her
the agency would own anything she wrote. Now that she was a
national celebrity, Christa's space journal might be worth millions of dollars. Was it fair for her share to be zero?

Her mother took up some of the slack with a journal of her
own.

"One of her greatest concerns," Grace Corrigan wrote of her
daughter's early weeks in Houston, "was whether she would be
accepted by the other astronauts and their families."

The crew plus spouses and kids often gathered at the Scobees'
ranch house in Clear Lake, a subdivision just beyond the Johnson
Space Center's fence. When June Scobee, the commander's wife,
told Christa she'd better come to Clear Lake if she knew what
was good for her, payload specialist McAuliffe practically saluted.
The Scobees were America's space-age dream come to life, a kid
who dreamed of flying and the schoolgirl who fell in love with
him. June was the mission's mother hen, permed and petite at
five foot two—a foot shorter than her husband, who liked to say
she was the brains of the family.

Christa and Barbara Morgan appreciated being invited to the

shuttle crew's "happy hour buffets," as June called them. The commander's ranch house featured model airplanes and a small scale model of the shuttle. When the Scobees first moved in they couldn't afford art for the walls, so Dick taught himself to paint and produced portraits of his favorite subjects, a T-38 Talon and an F-4 Phantom fighter jet. June stitched a wall hanging that featured a brilliant sun and Einstein's equation $E=mc^2$. The Scobees were comfortably settled parents of two by the time the "new girls" came by to get acquainted with the *Challenger* crew. There was mission specialist Ron McNair, whose wife, Cheryl, was a school-teacher like them. Dr. McNair, a laser physicist, had earned his golden pin the year before, becoming the second African American to fly into space. A gifted saxophonist who gigged around Houston, he had played a miniature sax in orbit and planned to record an original tune during their upcoming flight. The best touch football player in the crew—a little ahead of pilot Smith, who had been a star quarterback in high school—McNair was also a martial arts instructor, a black belt who could smash concrete blocks with a single karate chop.

Mission specialist Ellison Onizuka, the joker in the crew, liked to razz the commander about his grill skills. "Nice work," he'd say when Scobee dropped a burger into the coals. A Hawaii-born air force pilot, Onizuka had flown on *Discovery* in January 1985, becoming the first Asian American astronaut and first Buddhist in space. He and his wife, Lorna, and their daughters lived near the Scobees in Clear Lake, where Onizuka was known for hosting luaus complete with *kalua* pork prepared the old-fashioned way. He would roast a whole hog overnight in an underground pit lined with rocks, wood, and leaves, until the pork was so tender it fell off the bone. Despite his years as a test pilot and lieutenant colo-nel in the air force, Onizuka joked that he was a proud beneficiary of NASA's diversity program. "If I didn't have these slanty eyes, they never would have picked me." Dick Scobee and El, as he was known, did yard work together, trading tools, flight-school memories, and the occasional barb. "This guy can command the most complicated machine ever built," Onizuka said, "but a spatula? Look out below."

Mike Smith, who would sit to Scobee's right as second in command, was more of a straight arrow. (His title had a history of its own, as astronaut Bob "Mr. Shuttle" Crippen explained: "We use the terms 'commander' and 'pilot' to confuse everybody. It's really because none of us red-hot test pilots wanted to be called a co-pilot." The driven flyers who made astronaut bridled at being anybody's "co-" anything, so the agency dubbed shuttle pilots "commanders," allowing the seconds in command to be called pilots. "In reality, the commander is the pilot and the pilot is a co-pilot.") A jut-jawed navy flyer, Smith was about to turn his astronaut pin from silver to gold—their mission would be his first spaceflight. Like Scobee, he had flown in combat during the Vietnam War. But while the commander flew cargo ships, Smith piloted Intruder jets on air strikes over North Vietnam, turning back at 600 miles per hour to land on the USS *Kitty Hawk*, putting his wheels down at 150 miles an hour with a hundred yards of deck between him and the South China Sea. Smith was awarded nineteen medals, including a Distinguished Flying Cross. Now forty, with two daughters playing tag in the Scobees' backyard, he was the kind of flyer whose pulse didn't vary whether he was taking control of the shuttle in a simulator test or breaking up a tiff between his girls and the other kids. A military man drilled from adolescence in the chain of command, Mike Smith wouldn't dream of teasing his commander the way Onizuka did.

Judy Resnik could explain the aerodynamics of a Frisbee while diving past the flyers to catch one. While some of the other women enjoyed white wine, she drank beer like the boys. In addition to her PhD in electrical engineering, Christa's science mentor was a licensed pilot and classical pianist who had scored 1600 on her SATs, one of fifteen perfect scores recorded up to that time. In interviews she insisted she was "not a Jewish astronaut or a lady astronaut. I'm an *astronaut*." Christa, an overachiever in her own right, tried not to be overawed by Resnik and some of the other astronauts she was getting to know. She knew they were a special breed. They had all overcome odds as long as the 11,000-to-1 chance that propelled her in the Teacher in Space competition. Astronauts were and are a subset of Americans representing .0000005 percent of the population, a select group of death-defying valedictorians. Even in that company, Resnik stood out.

"Judy was a strong personality," says June Scobee. "She was determined to be as good as any astronaut and better than most. She was gorgeous, as everyone told her, not that she liked hearing it. She was very smart and very, very competitive—they all were. When somebody told Judy that women couldn't do this or that, she'd say, 'Watch me.'" Resnik was the only unmarried member of the *Challenger* crew and the only one with an unlisted phone number. She was also the only one with a stalker. According to her fellow crewmate Mullane, a young engineer who'd worked for one of NASA's contractors "went, quite literally, mad for her." The fellow sent gifts, proposals, and letters, including a poem hailing Resnik's "raven hair and eyes." He sneaked into a military-only area where she was about to take off in a T-38, pulling the chocks from under her wheels so she could taxi to the runway. One day he showed up at her desk in the Astronaut Office, grabbed her arm, and led her down the hall, saying, "We've got a meeting." Security guards intercepted them and took the stalker away. She never heard from him again.

Other astronauts called her "J.R.," a tip of the Stetson to J. R.

Ewing, the oil-mogul antihero of the TV show *Dallas*. To Christa, Resnik was just about everything an American woman could be. It was strange to think that she would have been one of the subjects of her American Woman class if she had flown before Sally Ride instead of the other way around. Now they were becoming friends. And something better: crewmates. Of all the flight-training stories Christa heard that fall, one of her favorites had to do with the time J.R. entered the astronauts' locker room while a male colleague, one of the *muy macho* ones, was naked—and *he* blushed. All she said was, "Nice butt, Tarzan."

Smith, Onizuka, McNair, Resnik, Jarvis, McAuliffe—it was the commander's job to turn this disparate bunch into a unit he could count on to deploy two satellites, perform more than a dozen in-flight experiments, and save each other's lives if they had to.

Sometimes they met at a restaurant on NASA Parkway. Despite its name, Frenchie's was an Italian joint where the walls held photos of shuttle crews and space walks, less crinkly than the older memorabilia on display at the Outpost. Scobee sat at the head of a long table, dispensing dinner orders while Onizuka reminded him that any pizza without pineapple might be cause for mutiny. Resnik, a long-distance jogger with a hearty appetite, reached for garlic bread, while rookie pilot Smith peppered her and the other spaceflight veterans with questions about their flights. Frenchie's proprietor Giuseppe Camera called Scobee "El Comandante" and hailed Resnik as *mia bella mora*, "my beautiful blackberry." Jarvis and McAuliffe sat at the far end of the table, befitting their station as payload specialists, until Scobee traded places with them. "Dick was proud and happy to be named their commander," June Scobee recalls. "That role was so important to him. He made sure they bonded as a group." The Scobees saw

Dick's first command as the ideal payoff for a quarter century of hard work and high hopes.

——◆——

Francis Richard Scobee, the first enlisted man ever to make astronaut, started out as the son of a railroad engineer in Auburn, Washington, half an hour from Seattle. Working bean fields to pay for his school clothes, young Scobee grew tall and strong and discovered that he had an aptitude for science and math. In 1957, he enlisted in the air force. He scored high enough on the entrance exam to be offered an entry-level job in military intelligence. He turned it down—he wanted to fly. The air force sent him to Kelly Air Force Base in San Antonio, but not as a pilot. He was an airplane mechanic. In his free time he took engineering classes at San Antonio College.

One night, the lanky young mechanic joined a church-sponsored hayride to a farm where the stars were so bright you could almost reach up and touch them. Another hayrider caught his eye. Sixteen-year-old June Kent was as curious about the wide world as he was, and sprightly and cute as a bug. Scobee was smitten. After the ride back to Mayfield Park Baptist Church he offered to drive her home, forgetting that his 1950 Chevy had ignition trouble. She had to help push before they got it to start.

His "Junebug," as he called her, came from Odenville, Alabama, where her carpenter father went bankrupt. Her mother ranted at shadows and spent months in a mental hospital. Schoolchildren jeered that her family was "so poor they stink, they got no kitchen sink!" But June was a born optimist. After they married—he was nineteen, she was sixteen—she said he ought to apply to the Air Force Academy. "You still want to fly, don't you?" But Dick only shrugged. "I'm not what they're looking for."

Without a recommendation from a congressman he wouldn't have a chance, he said, and his family had no political connections. "So here I am, a mechanic at an air base, which is as close as I can get to flying a plane."

They both took night-school classes at San Antonio College. June took time off to have a son, Richard, in 1957, and a daughter, Kathie, in 1961, four months before Alan Shepard became the first American astronaut. Her husband wielded a wrench at the base and took enough college classes to earn a bachelor's degree in aerospace engineering. In 1965, the air force assigned him to pilot training at Edwards Air Force Base, in California's Mojave Desert, a grid of hangars and airstrips on a vast, ancient lake bed where Chuck Yeager had piloted the rocket-powered Bell X-1 that broke the sound barrier in 1947. It was Yeager, the original avatar of the Right Stuff, who said, "Progress is marked by great smoking holes in the ground."

Dozens of astronauts earned their test-pilot wings at Edwards. Eleven went on to die in the line of duty. Soon after Scobee's arrival, with the Vietnam War heating up, the air force gave the twenty-six-year-old grease monkey his first chance to fly.

"He was a natural," June Scobee recalls. Instructors gave the strapping young pilot high marks and their highest compliment: *He's a good stick.* He could fly single-seaters as easily as the C-141 Starlifter, a 170-ton cargo plane that handled like a whale. Scobee's skill was a matter of proprioception, the inner sense that guides our movements. "He could feel every part of a plane as if it was part of himself," June says. Dick Scobee with his Brylcreemed crew cut could feel a plane from nose to tail fin as if it were part of his nervous system. For two years he flew C-141s through antiaircraft fire over North Vietnam, winning a Distinguished Flying Cross.

"I remember Dick from Vietnam," another pilot recalled. "He

saved my life when he kept the enemy at bay while I belly-landed my fighter." Scobee never mentioned that flight to his wife and children.

After the war he flew prototypes of the Boeing 747 and the space shuttle, offering insights on their handling and aerodynamics to the crafts' designers. In the 1970s, he grew out his crew cut and looked "handsomer than ever," his wife says. By 1985, Scobee was one of the space program's esteemed veterans, which made him a target of trash talk about his age. "Old man, sir," the thirty-nine-year-old Onizuka would call him, or "Grandpa, sir," suggesting an experiment on the effects of weightlessness on an old man's bones and saggy testicles.

Still lean and fit at forty-six, Scobee didn't have to prove himself. Everybody in the Astronaut Office knew he could land a paper airplane in a hurricane. Payload specialist Greg Jarvis remembered a training exercise:

"We were in the Motion Base Simulator when the lights went out on the visual for landing," Jarvis recalled. To make training as realistic as possible, the simulator's windows featured projected images that matched what the crew would see in flight. That day, with Scobee at the controls, the flight-deck windows went dark. He asked their trainers, who were acting as Mission Control, "Are the lights out?"

As Jarvis recalled, "They said, 'We'll get back to you on that.' This went on for about two or three minutes, and it turns out they had mistakenly turned the lights out on the visuals." Which troubled Scobee not at all. "He made a perfect landing without any lights!"

Comparing *Challenger* to the rocket-powered plane Yeager flew through the sound barrier in 1947, Scobee said, "It's not like the old days anymore, when you'd jump in a plane and take off. What we do is more analytical." As the *Washington Post* put it, "The new breed of test pilots looked back on *The Right Stuff* days mostly with awe, but also with a little condescension. There seemed to be a New Stuff, and Dick Scobee was its perfect embodiment."

His wife became the first in her family to get a college degree. June Scobee earned her PhD in education in 1983 and went on to direct the space science program at Texas A&M. The two of them discussed the risks he took with every mission. "We really wrestled with life and death," she says. "Was it worth the risk? To do what he'd dreamed of since he was a little boy, and more—to fly into space? We said yes, and after that every day seemed like a gift. Every day was an extra day."

Asked if he ever felt fear on the job, Scobee said yes. "You would be stupid *not* to have a healthy fear. There's six and a half million pounds of thrust under you. But you have to risk something to gain anything."

Kathie Scobee turned twenty-three the year her father earned his gold astronaut pin. "His business was risky and technical," she recalls, "with a dash of instinct. He used to give human attributes to things. He'd pat the wheels of one of his planes and say, 'New shoes.' He'd come home from work and tell us, 'The computers weren't talking to each other today. They argued and made us crash.'" She remembers her dad as "a sweet, funny person with nothing negative in his heart. Well, most of the time. In public he'd put on his military look. 'Don't mess with me.' And he hated waiting in lines. We'd be at an amusement park and he'd grit his teeth and say, 'Patience, my ass. I'm gonna punch a hole in something!' But the way he loved my mom was beyond amazing. They got married at sixteen and nineteen and pretty much raised each other."

The agency chose Scobee to command the Teacher in Space

flight partly because his wife was an educator. "They want to assign the teacher to my flight," he told June when the Teacher in Space program was starting. "They think I'll be nice to a schoolteacher because I'm married to one!"

Six-year-old Caroline McAuliffe had her doubts. From her mother's descriptions of Commander Dick Scobee, Caroline pictured the man as a giant. During a family trip to Houston, where Christa introduced her family to the crew, Caroline clutched her dad's leg outside Scobee's office. "If he's scary," she said, "I'm not going in."

But Scobee wasn't scary. He told the McAuliffes about family life around the space center, where guys like him lugged their space helmets and samples of freeze-dried space food to show-and-tell days at their kids' schools, and the kids could not be less impressed. "They're like, 'Oh no, not another astronaut!'"

Scobee told Christa she was the most important member of his crew. "Shuttle missions are taken for granted these days. You're helping us change that. You're the reason we're going to be back on TV," he said. "No matter what happens, our mission will always be remembered as the Teacher in Space mission, and you should be proud of that. *We're* proud of it."

He showed them around his cramped office with its family photos, diplomas, and citations. There was a framed copy of "High Flight," a 1941 poem by pilot John Gillespie Magee:

> *Oh! I have slipped the surly bonds of Earth*
> *And danced the skies on laughter-silvered wings;*
> *Sunward I've climbed, and joined the tumbling mirth*
> *of sun-split clouds,—and done a hundred things*
> *You have not dreamed of—wheeled and soared and swung*
> *High in the sunlit silence. Hov'ring there,*
> *I've chased the shouting wind along, and flung*

My eager craft through footless halls of air. . . .
Up, up the long, delirious, burning blue
I've topped the wind-swept heights with easy grace
Where never lark nor ever eagle flew—
And, while with silent lifting mind I've trod
The high untrespassed sanctity of space,
Put out my hand, and touched the face of God.

Magee was flying for Britain's Royal Air Force when he crashed and died, three months after writing "High Flight." He was nineteen years old. Christa told Scobee she might start teaching the poem in her English class when she got back to Concord High School. He thought that sounded like a good idea. Till then, he said, she was going to have to work harder than she ever had before. "I'm counting on you," Scobee said.

Beyond that, he had one more order for her. "Once we get up there, look out the window every chance you get."

—◆—

After a month of training, Scobee wanted Christa to feel the G forces that would press her to her seat on the way up. He treated her to a ride in a plane that was faster than Chuck Yeager's X-1: a Northrup T-38, one of the agency's supersonic jets. The needle-nosed T-38, a two-seater with a Plexiglas bubble top over the cockpit, white with sky-blue trim and the NASA logo on the tail, topped out at Mach 1.3. Supersonic training flights were known as "zoom-and-booms." Christa's first spin in a T-38 was subsonic but fast enough to pin her to her seat. Belted into the seat behind Scobee, breathing through an oxygen mask, she whooped when he spun them through corkscrews at eight hundred miles an hour. As Scobee explained later, "The primary purpose was to put her in a stress environment, a situation she's never been in before.

Speed, helmets, confined to a cockpit. It's not your everyday air-plane ride, and if it bothers you, you probably are not ready for space."

To Christa's delight, this was nothing like her stomach-turning ride in the Vomit Comet. She laughed, gaping at the green-and-brown landscape five miles below—or above, as it appeared when Scobee rolled the plane and they were upside-down. Still, she got rattled when he drawled, "She's yours." He hadn't warned her about that. Fifteen thousand feet over Texas, he let Christa take the controls. Of course he could take them back in an instant if she made a wrong move. But for the next ten minutes, with Scobee's calm voice in her headphones, telling her what to do, Concord High's Mrs. McAuliffe—whose top technical feat to that point was beating several levels of *Ms. Pac-Man*—flew a million-dollar supersonic jet.

Dear Eileen,
The T-38's a high altitude jet and Dick let me have the controls and do rolls and dives . . . We also felt 3 Gs—it's like you are collapsing into yourself. . . .

Love, Christa

Her NASA handlers had let a photographer mount a small camera in the cockpit. During one barrel roll, Christa snapped a selfie. The photo shows her in the cockpit behind Scobee, the Gulf coast of Texas hanging sideways over her shoulder. "My greatest thrill," she called it, "so far."

6

THE *CONCORD MONITOR*'S BOB HOHLER TURNED UP AN OMI-
nous precedent for Christa's flight. In 1927, four days after Charles
Lindbergh's solo flight across the Atlantic, pineapple mogul James
Dole offered twenty-five thousand dollars to the first airplane to
make it from California to Hawaii. The Dole Air Race captured
the nation's attention, thanks in part to schoolteacher Mildred
Doran, who flew as a passenger in a biplane named after her.

"A woman should fly just as easily as a man," she announced
before climbing aboard the *Miss Doran*. "Women certainly have
the courage and tenacity required for long flights." As for the risks
of flying across 2,400 miles of open water, she said, "Life is nothing but a chance."

That August, eight propeller planes took off from Oakland.
Two reached Hawaii safely. Two crashed on takeoff, one turned

back, and three others, including the *Miss Doran*, disappeared somewhere short of Honolulu. Mildred Doran was never seen again.

Christa wasn't fazed by stories like Doran's. Aviation had made great progress in the fifty-eight years since the Dole Air Race, she figured. NASA had made giant leaps in the eighteen years since Apollo 1, its last fatal accident, advancing from shooting three-man capsules atop rockets to launching a vehicle that sat seven crew members, climbed like a rocket, orbited like a satellite, and landed like a glider. Asked about the hazards of riding on *Challenger*, she cited the shuttle program's unblemished safety record. "I don't see it as a dangerous thing to do," she said. "Well, I suppose it is, with all those rockets and fuel tanks. But if I saw it as a big risk, I'd feel differently." She compared shuttle missions to flying on TWA or Northeast Airlines or, worse, driving in New York City traffic. She agreed with Resnik, who once dismissed a reporter with four precise sentences: "It does not enter any of our minds that what we do is dangerous. The world might think it is. We don't. Something is dangerous only if you're not prepared for it, or you don't have control over it, or you can't think through how to get yourself out of the problem."

Christa's immediate dilemma was extracurricular. She had promised to write up a favorite recipe for a *Teacher in Space Cookbook*. Along with Cloud Nine Cheese Puffs and Shuttle Salad, the book would feature President Reagan's favorite macaroni and cheese, John and Annie Glenn's ham loaf, and Dick and June Scobee's M&M cookies. Christa had a hard time choosing between her broccoli casserole and a New England specialty, Spicy Apple Pancakes with Cider Sauce, so she contributed both.

Back in Concord, Steve McAuliffe joked about writing a tome of his own, *The Space Husband's Cookbook*, with a chapter on "Snickers, the Forgotten Breakfast Food."

Steve was sober and direct in court, but he had a flip side. When a reporter asked if he and his wife were Democrats, he asked how anyone with a conscience could be a Republican, then added, "Say I said that with a twinkle in my eye." His best-known case had pitted the ACLU against New Hampshire governor Meldrim Thomson, who ordered flags on state buildings to fly at half-staff on Good Friday. Steve and his partner Bill Glahn, representing the governor, took their case all the way to the US Supreme Court. With less than twenty-four hours to file appellate briefs, they wrote arguments on legal pads during a flight from Concord to Washington, DC, then raced to the Supreme Court Building. The high court's clerk turned them away, saying, "The Supreme Court does not accept handwritten briefs."

Steve threatened to take his case "to a higher court!" even though there wasn't one. He won that battle—the clerk accepted their briefs—but lost the war. "Working hastily from handwritten legal papers," the *New York Times* reported, "the high court voted, 5 to 4, to reinstate a lower court order barring Governor Thomson from carrying out his proclamation."

Seven years later, Concord's most prominent young attorney was amused to become America's favorite househusband. "Describing himself as the perfect star for a sequel to *Mr. Mom*, McAuliffe is four and a half months into his stint as a single parent," the *Monitor* reported. "Only Mary West, a cleaning woman who visits two mornings a week, stands between him and domestic disaster." While Christa trained in Houston, Steve said he was starting to understand how England's Prince Philip felt. Christa's "happy hubby," as the *Boston Herald* described him, charmed reporters with tales of his efforts to keep two kids fed, dressed, washed, and uninjured. He said he'd be up a creek if not for friends who dropped by with lamb chops and buckets of Kentucky Fried Chicken. He told reporters about a magazine quiz he and Christa had taken called "Who Knows What Around the House." Final score: Christa 94, Steve 6.

Like Dick Scobee and Mike Smith, Steve had grown up building model airplanes, hoping to be a jet pilot. "Flying was all I ever wanted to do." Now his wife was preparing for the twenty-fifth space shuttle mission while Steve did the laundry. His trajectory was a lesson in the absurdity of expectations, and more. It was a thrill. "The best part of all this," he said, "is how much better I've gotten to know our kids."

——◆◆——

Payload-specialist training consisted of 114 hours of classwork, plus simulations and drills known as "familiarizations." The simulations were called "sims," the familiarizations "fams." NASA described the program as "a comprehensive flight training course" that would make trainees "familiar with shuttle systems, payload support equipment, crew operations, housekeeping techniques, and emergency procedures."

As hard as the agency tried to control every variable, no system is perfect. "On my first flight, I knew we were weightless when a piece of cookie from the previous mission floated up in front of my face," one astronaut remembers. "Followed by a stray bolt that had gotten loose from someplace. I only hoped the bolt wasn't the one that was supposed to hold the whole orbiter together."

Housekeeping matters on the middeck included operating the food and microwave compartments in the galley as well as the vacuum-powered space toilet. As one astronaut noted, "Juice allowed to drift from a container would turn into a ball that would float until it hit a wall and divided into two spheres, leaving a sticky mess the crew would have to clean up." The same was true of any stray droplet of urine.

According to the agency, "payload specialist training may

begin as much as two years before a flight." Christa had less than five months to train for her mission, with constant interruptions for press junkets.

In October, she climbed aboard the Vomit Comet to rehearse the zero-G demonstrations she would conduct from space. Remembering Resnik's advice, she bit her ScopeDex tablet in half, and this time she enjoyed her two-hour ride in the padded plane. She and Barbara Morgan played catch with a weightless pill bottle, did a floating Cossack dance, and went leap-frogging through the cabin. "The first leapfrog in zero gravity!" Christa called it.

Later that month, the agency flew the two of them to Florida for a shuttle launch. On a sunny 75-degree day at Cape Canaveral, she and Morgan watched *Challenger* lift off on its last flight before the Teacher in Space mission. "Oh, my god," Christa said, laughing with pleasure as the shuttle rose from the launching pad in clouds of flame and steam. "It's beautiful!" Later, looking forward to her own liftoff, she told reporters, "I've talked with my kids about the launch. Some astronauts' kids have had bad reactions, but I think they'll be okay. Caroline is six now and Scott's nine. They're aware of rockets and how they behave. Caroline doesn't like loud noise, so I told her it can be loud." Asked if she considered the Space Flight Participant Program a success, she said yes. "I'm hoping everybody out there who decides to go for it—the journalist in space, the poet in space, and the other categories—that you'll push yourself to get the application in."

As astronaut Mike Mullane remembered, "There were rumors Walter Cronkite and John Denver were being considered for flights." The rumors were true. After making Christa a celebrity, NASA invited reporters and TV personalities to apply. One TV star the agency seriously considered for the payload-specialist program—*Sesame Street*'s eight-foot-tall Big Bird—was disqualified because he was too tall for the crew cabin. But poets and artists were under consideration, and journalists were next in line after the Teacher in Space. As NASA Administrator James Beggs recalled, "We had been thinking of taking a journalist up for a long time, because we thought if we could get a journalist up there, somebody would give us, if not good publicity, at least a lot of publicity." In the end, 1,703 applied to be the first Journalist in Space. They included CBS News anchor Walter Cronkite, NBC anchor Tom Brokaw, and ABC White House correspondent Sam Donaldson. The sixty-nine-year-old Cronkite said he wanted "to prove that any old fart can do it." Brokaw, quoting every pilot's favorite poem, asked, "What more could you ask than to slip the bonds of Earth?" Geraldo Rivera's application posed another question: "To beat through the air and clouds and sail through the vast ocean of vacuum, what must that be like?" As *Newsweek* columnist George Will put it, "They are probably taking a journalist on the principle that Earth could not but be improved having one fewer on it." Another hopeful was Norman Mailer, who would turn sixty-three in the week of Christa's launch. "Obviously I'd like to go," Mailer told *Life* magazine. "I am, however, making no attempt to get into shape since I do not have the chance of the honorable snowball in hell. When it comes to taking that kind of chance on a writer, NASA has about as much imagination as Nabisco. My apologies to Nabisco."

The agency winnowed the journalists' entries down to forty semifinalists. Cronkite made the cut, along with *New York Times* science writer John Noble Wilford. Astronaut Mullane doubted that the candidates knew what they were getting into. "The entire part-timer program," he claimed, "was built on the lie that the shuttle was nothing more than an airliner which just happened to fly higher and faster than a Boeing 747. The very act of assigning a schoolteacher and mother of two to a shuttle mission dramatically reinforced that lie. But every astronaut knew what the shuttle was—a very dangerous experimental flying rocket without a crew-escape system."

In order to overcome Earth's gravity the shuttle had to ride piggyback on an enormous fuel tank and two silo-sized rockets. It had to be armored to withstand a fiery reentry and streamlined enough to glide to a landing. As astronaut Story Musgrave put it, designing such a contraption was "like bolting a butterfly to a bullet."

But it worked. As of Halloween 1985, the shuttle program could boast of twenty-two launches and twenty-two safe landings.

—◆—

In the *Monitor*'s back-to-school supplement, "A Student Guide to the Space Shuttle: Christa's Challenge," she fielded students' questions:

> Q: *How does it feel to suddenly be a celebrity?*
> Christa: *It's fun. The media experience is new to me, but I'm enjoying meeting lots of people and getting an opportunity to see "behind the scenes" at TV and radio stations.*
> Q: *Are you having second thoughts about doing this?*
> Christa: *I get more excited about the trip as I do more of the training with the crew. I'm sure the adrenalin will be flowing when I'm blasting off!*

The longer she trained, the more she felt a kinship with Greg Jarvis, the mission's other payload specialist, who became a friend. They were partners in sims and fams, as well as in the outsider role they shared.

It would have been easy to underestimate the forty-one-year-old Jarvis. At five foot nine, he barely came up to the commander's ear. With his thinning hair and less-than-chiseled physique he could have passed for a Little League coach, but few coaches could match his athleticism. He and his wife, Marcia, lived in Hermosa Beach, California, where Greg drove his rusty Dodge Dart to the beach to go surfing before work. A runner, long-distance cyclist, and whitewater rafter, he had beaten out six hundred other Hughes Aircraft engineers to win a spot on the shuttle *Discovery* only to be ousted by Senator Jake Garn. Reassigned to *Columbia*, Jarvis got bumped again, this time by congressman Bill Nelson. He was re-reassigned to the Teacher in Space mission, which didn't have a Hughes satellite for him to deploy. "For Hughes," the *Washington Post* reported, the mission was "a combination of public relations and unprecedented firsthand observations." For Jarvis it began with months of rehashing his training for *Discovery*. Sixteen years after earning his master's degree in electrical engineering, he spent free hours working on a master's in business management, "just to keep my mind active."

Compared to Christa, Jarvis was a veteran at JSC, but he still had trouble in the pressure chamber, where test subjects breathed through oxygen masks while instructors thinned the air until it matched the pressure at thirty-three thousand feet, almost a mile higher than the summit of Mount Everest. Then the trainees were told to remove their masks. "At that altitude, your brain doesn't get enough oxygen. You'd die in about two minutes," said Hughes Aircraft engineer Steve Cunningham, who was Jarvis's backup on the *Challenger* mission. "They had us take our oxygen masks off, then gave us a clipboard with questions on it. Greg was very intent on answering the questions. One instructor who realized he was past the point of no return took the pencil, turned it around with the eraser down, and put it back in Greg's hand. He didn't even notice. The instructor shook him and told him to put on his oxygen mask. Greg just smiled. The instructor grabbed the mask and put it on for him. We realized that this was no game, this was serious."

Christa aced the pressure-chamber test but fumbled some of

her lines in rehearsals for her space lessons. So she put her script aside. As Morgan remembered, "She said, 'I'm not an actress. I'm not pretending to be a teacher. I *am* a teacher, and teachers don't need speeches. All we need is a lesson plan.'" After that she improvised with tips from Resnik and Morgan. Before long Christa was confident enough to suggest camera angles for her lessons. During one rehearsal in the crew compartment trainer, she climbed down the ladder from the flight deck, saying, "The camera should pick me up coming down." Describing an in-flight exercise routine, she said, "As I'm talking about the treadmill, this is the cue for the crew to start floating into view." After knocking a plastic bottle to the floor during one rehearsal, she joked about it. "Oh, this zero-gravity environment is awful!"

One of her clashes with the agency involved a set of lesson plans that would go out to classrooms all over the country. The sixteen-page "Teacher in Space Teachers' Guide" had been drawn up by a NASA writer whose introduction claimed that Christa had worked hard to prepare it. In fact she'd had nothing to do with the project and thought it reeked of condescension (*"Have students work individually or in small groups. . . . Ask students to think about their interests"*). The guide misspelled Judy Resnik's name (*"Mission Specialist Resnick is a classical pianist"*). It presumed that busy teachers would devote weeks of class time to prepping students for Christa's fifteen-minute transmissions from *Challenger*. She used a red pen to cross out sentences and whole sections of the guide. Her NASA editors dutifully removed the invented lines about her slaving away on the text but left the rest intact. They made plans for her to present the guide to the president of the National Education Association at an upcoming photo opportunity. Christa refused. Her NASA bosses asked Morgan to do the honors instead. Morgan refused. As the Associated Press eventually reported, "Mrs. McAuliffe was not on hand as the lesson plan was presented to the National Education Association's president, Mary H. Futrell. That is because she is training at the Johnson Space Center in Houston."

She was learning to use the power her position gave her. One morning she and Morgan were running late to a press conference, hurrying around a blind corner at JSC. Christa laughed and said, "Don't worry, they can't start without us."

Still she picked her fights with care. When the White House lodged a complaint about Christa McAuliffe, she listened.

As her mother noted, "Christa was a vocal New Hampshire Democrat. She admired Eleanor Roosevelt and the martyred Kennedy brothers." After the *Boston Globe* described Christa as "an outspoken union activist, a self-described feminist and Kennedy Democrat" and quoted her praising President Kennedy, NASA Administrator James Beggs got a call from Michael Deaver, one of President Reagan's advisors. "Deaver told me to tell her not to talk," Beggs recalled. "I told the folks down at the Astronaut Office to say, 'Look, Christa, we can't tell you not to talk to the press. But recognize the fact that you are in a very sensitive position politically and try not to be partisan, because this is a nonpartisan agency.'"

This time she listened. From then on, she gave reporters all the quotes they wanted about the mission, but none about politics.

Dear Eileen,
I'm still enjoying this different life. Thank God for Barbara though. We both agreed that if we were doing this solo we wouldn't last. We get to relax and put things back in perspective when we are away from "the center." . . . I'll try to call you at school.

Eileen O'Hara made way for another substitute in room 305 that fall. Secretary of Education William Bennett, making good on his pledge to fill in for Christa for a day, strode into Concord High one morning in a cream-colored suit, trailing a crew of aides and journalists. "A small circus," the *Monitor* called Bennett's cameo. There were more reporters than students present as Reagan's education chief launched into a lecture on the Federalist Papers.

"It was kind of a joke," recalls Kevin Swope, a Concord High senior that year. "The school had put three ringers in the class"—honor-roll students who might make a good impression on the secretary. "I was one of them." So was his irreverent classmate Mike Bart, who had annoyed Christa during one of her field trips to Washington, DC. On that trip she waited up half the night, caught him sneaking into the hotel, and gave him a lecture about selfishness. Bart liked to joke that he was the only person Christa McAuliffe ever hated.

"We were advised not to say much in Bennett's class," Swope says. "Certainly nothing controversial." But Bart had read up on the secretary's record. Bennett opposed affirmative action. He supported school prayer and cuts in student-loan funding. When Bennett took a breath, one of the ringers raised his hand. "Of course it was Mike," Swope recalls. "He asked, 'How come the Reagan administration keeps cutting funds to education?' Bennett hemmed and hawed, then went back to telling us about the Federalist Papers."

After the class, Secretary Bennett and Principal Charles Foley shook hands and smiled for photos. Bennett and his aides headed for the airport, and Foley returned to his duties, which included making sure the school had enough AV carts to roll a TV set into every classroom for the launch.

In Framingham, Grace Corrigan subbed for her daughter as a special guest at Framingham State College's 1985 homecoming. By then the Corrigans had hung their daughter's official NASA portrait in their living room. Christa had sent her mother a greeting card showing a woman climbing a mountain. *"Behind every great woman . . ."* it read. Grace opened the card to read *". . . is another great woman. THANKS MOM!"* Fifteen years after marching to protest the Vietnam War, Christa was a national hero, her picture all over the Framingham State campus. That year's homecoming theme was "It's Out of This World."

At Kennedy Space Center in Florida, *Challenger* lifted off for its penultimate flight on October 30. Commanded by Hank Hartsfield (who had locked the hatch to foil a payload specialist on a previous flight) and carrying Guion Bluford, the first Black astronaut, on his second mission and Bonnie Dunbar, the seventh female astronaut, on her first, the shuttle circled the globe for a week before landing at Edwards Air Force Base on November 6. Then *Challenger* was inspected for damage, picked apart for components to be used on other missions, and flown across the country, riding atop a modified passenger jet, for two months of preparation at Cape Canaveral before the Teacher in Space mission.

At JSC, in Houston, Christa practiced using the equipment and supplies she would deploy in orbit. Everything she would need for her televised lessons—her own specialized payload—had to fit into a locker that measured seventeen inches by fourteen by twenty-two, about half the size of a school locker. Her hydroponics lesson was to involve white beans planted in nylon mesh instead of soil or pebbles, but a pre-experiment experiment proved that mung beans worked better. A demonstration of Newton's laws would feature a billiard ball and a smaller steel ball to demonstrate inertia and the action and reaction of the balls' colliding. Christa's practice for a demonstration of magnetism started with iron filings and a small electromagnet, but iron filings might get loose in the cabin and pose a threat to astronauts, who might inhale them, so her trainers switched to short strands

of wire—and made sure the electromagnet was too weak to inter-
fere with the shuttle's instruments.

Her trainers also continued preparing detailed scripts for her
broadcasts from space. The first, discovered years later, reads:

THE ULTIMATE FIELD TRIP/LESSON 1

(2 camera setup/3 scenes)
T= 15:00
*Scene opens on aft flight deck (extreme wide) with Christa
          and CDR Scobee front and center.*
*Christa will introduce her lesson theme and begin with an
          orientation to the flight deck . . .*
*During scenes 2 & 3, MS Resnik will support the lesson with
          camera zoom, when necessary,
          from the aft flight deck. Ends with: Q & A*

"She's been poked and probed," *USA Today* reported. "She has
learned to stand on her head in zero gravity and been subjected
to surprise tests to see if she'd panic under pressure." Between
tests, Christa baked the crew a cake. The others raved about her
cooking. Scobee praised her to the press. "She's got a positive attitude," he said, "and stays out of the way when she needs to stay
out of the way."

Like any payload specialist, she hoped to prove she belonged
on a shuttle crew, but could never be sure what the career astronauts might be saying behind her back. As her mother wrote,
recalling what she called "Christa's fears" in Houston, "Her role
was unique—she was the first private citizen to infringe on the
astronauts' tightly knit core group." And Scobee was protective of
his core group. When a *Life* magazine photographer told Christa
to jump off a desk into the commander's arms, Scobee wanted no
part of it. "Take Christa's picture. She belongs to the nation," he
said. But if *Life* or anyone else wanted a picture of him, it would
have to include the whole crew.

Each crew posed for an official portrait that would hang in Houston, at Cape Canaveral, and at NASA headquarters in Washington, as well as at Frenchie's and the Outpost. When Christa arrived at the Astronaut Office for the latest photo session, she found the rest of the crew dressed in sky-blue flight suits. Scobee said they might not need her after all. He asked her to wait in the next room. She followed orders and spent several minutes waiting and worrying. Finally, Scobee came through the door, leading the others. They were grinning, dressed in shorts and short-sleeved shirts, carrying Cabbage Patch Kids lunch boxes and apples for the teacher. Resnik had a little girl's purse slung over her shoulder.

Scobee said, "Hi, teach."

7

TRAINING SPED UP IN NOVEMBER.

The Vomit Comet, with its brief spells of weightlessness, gave trainees a taste of zero G, but the padded airliner was no place for any drill or simulation that lasted longer than thirty seconds. NASA engineers had discovered that water could serve as a suitable stand-in for the vacuum of space; after that, they put shuttle crews through simulations in a swimming pool.

Per agency tradition, the pool in Building 29 at the Johnson Space Center could not be called a swimming pool. It was JSC's "Weightless Environment Training Facility," or WET-F. Twenty-five feet deep, attended by a pair of pneumatic cranes that lifted full-scale models of the shuttle's compartments and dunked them into the water, the WET-F was where Christa and her crewmates wore spacesuits underwater, coached by instructors in scuba gear.

One AsCan called it a fine substitute for zero G in every way but one: it wasn't much good for practicing extinguishing fires in the cabin.

Christa surprised herself by enjoying several spins in the "gimbal rig," a spherical device that whirled subjects in every direction to simulate the tumbling of an out-of-control space-craft. Other exercises took the crew outdoors—not only at JSC but at Ellington Field, ten minutes away, and in Florida, where the crew splashed through shuttle-escape drills at the air force's water survival school near Homestead Air Force Base, south of Miami. Christa was getting used to the idea that earthly distances were no obstacle to astronauts. Civilians might need airports to go from place to place, but shuttle crews could zip from Ellington to Homestead or KSC in NASA T-38s and get back to Houston in time for dinner. At water survival school she parasailed behind a powerboat, smiling for camera crews. In another drill she zip-lined down a cable to simulate a parachute landing, dropping into a life raft under a rescue helicopter, laughing all the way.

Scobee scolded her for that. "Be serious," he said. "This can save your life."

Scobee recognized the absurdity of putting a space-shuttle crew through Gemini-era escape drills, but rules were rules. *Columbia*, the original shuttle, had carried ejection seats on its first missions, when the two-man crew consisted of the commander and the pilot. Ejector seats in a space shuttle were engineering Hail Marys in the first place. They might work during the few seconds just after launch or before landing, amounting to three or four minutes in a weeklong mission. Bailing out of a shuttle on a rocket-powered seat at any other time would mean suicide by freezing, asphyxiation, or the fires of reentry. Ejector seats served mainly to make astronauts feel the agency was looking after them, but larger shuttle crews on different decks made them still more impractical. After 1982, shuttles went without them.

With personal rescue spheres still in development, crew members knew there was little chance of escaping the vehicle if a crucial component failed. Still, they went through escape drills.

"The shuttle flies over the shark-infested Gulf of Mexico, so we had shark-survival training," an AsCan said. "We joked about who would be shark bait if the aircraft nosedived into the gulf while the rest of us scrambled to the lifeboat."

There were three space-shuttle simulators at JSC. Commander Scobee and pilot Smith mastered all three. They took the controls in the simulators' cockpits like an airliner's pilot and copilot, facing a bank of windows that provided a view of Cape Canaveral. Computer graphics allowed the simulators' windows to change with the shuttle's trajectory—except when the graphics conked out, as they did during Scobee's blind landing in one of the simulators. Hidden speakers blared the roar of engines and rocket boosters, the rustbucket creaks of metal joints straining, the humming of air vents, the beeps and bells of caution-and-warning alarms. The four crew members who would sit on the flight deck—Scobee, Smith, Onizuka, and Resnik—spent eight to twelve hours a week in the simulators, working their way through simulations that went smoothly and sims that didn't, leading up to intensive drills in the last six weeks of training. "We had fifty-six-hour sims," said astronaut Mike Coats. "They'd set up cots that you'd sleep in and go around the clock. They'd wake you up in the middle of your sleep with horns going off because a meteorite just poked a hole in the spacecraft and you had to do an emergency de-orbit."

The astronauts' coach during such sessions was the simulation supervisor or "sim sup," pronounced "sim soup." One of the sim sup's tasks was to throw curveballs at the crew. In some cases the sup would hand the commander a green note card scribbled with a particular crisis: *air leak* or *cabin fire* or *APU failure*. As training got more intense, the crew might get two or three green cards at once.

In 1983, when Guion Bluford became the first African American astronaut named to a shuttle crew, his mission's sim sup declared that the crew would have to contend with a *medical emergency*. The astronauts were left to decide what it was: A passed-out pilot? A payload specialist with appendicitis? As astronaut Mike Mullane remembered, "NASA had been in orgasmic ecstasy over the impending flight of America's first Black astronaut." After some hammy wailing and hand-wringing, the others relayed the nature of their emergency: "Guion turned white!"

There was one curveball no supervisor threw: the unsurvivable malfunction. "There are things up there that will eat your lunch, either in small bites or big gulps," one trainer said. The biggest gulps had no training value. "Our object is to push them to the limit of what they *can* get out of, and then push that limit so that it grows and grows and grows. It doesn't teach them anything to show them something they cannot do."

Still, a simulation had to test astronauts' skills. "We gave them malfunctions and contingencies," recalls Frank Hughes, NASA's chief of flight training at the time. Contingencies were worse than mere malfunctions. "One was a cabin leak. The shuttle architecture was such that if you had a hole in the wall a foot across, you're going to die, but if the hole is a half-inch across, the system was designed to feed the leak—it would pump oxygen and nitrogen into the cabin to feed the leak, to give you time to look around. If you could find the leak, you could stuff something in the hole." If not—or if you threw a wrong switch or made any other move that would doom the craft and crew—the sim sup stopped the simulation. Another contingency was the engine-out abort. During one of those, only near-perfect reactions would keep the simulation going. "In just about every one of those, we ended up in the Atlantic Ocean," an astronaut said. When Onizuka, the jokester of the *Challenger* crew, suspected they had an engine-out simulation on tap, he showed up in flippers and a scuba mask.

Christa wasn't needed in JSC's flight-deck simulators. She would have little to do during ascent and reentry, the brief, intense periods astronauts called "uphill" and "downhill." "Going uphill and down, payload specialists were cargo," says Hughes. During training, "we weren't teaching them to fly, we were teaching them to eat, sleep, and go to the bathroom" in space. Because payload specialist McAuliffe would spend almost all her mission time on the middeck, *Challenger*'s living room, her drills at JSC centered on the crew compartment trainer, a full-scale replica of the shuttle. At 170 square feet, about the size of the McAuliffes' kitchen in Concord, the CCT's middeck held storage lockers for the crew, a microwave oven, food trays, hot- and cold-water dispensers, a treadmill, and zero-G sleeping bags, as well as the device her Concord High students were most curious about, the shuttle's vacuum-powered waste collection system. High schoolers enjoyed learning that the WCS had stirrups and handholds to keep astronauts from flying off the toilet due to the equal and opposite force of their expulsions. The students liked hearing that according to Sally Ride, using the space toilet felt like "sitting on a vacuum cleaner." Equipped with twelve high-pressure jets under the seat, it worked by creating a partial vacuum that sucked waste matter into a compartment featuring a fanlike device called a "slinger" that flung solid waste to the periphery, to be freeze-dried for post-flight collection. When the slinger went haywire during one shuttle flight, releasing unmentionable particles the astronauts had to capture in baggies, they joked that this was the moment when the space shit hit the fan. They were glad it was their last day in orbit.

—◂◆▸—

While Christa trained in Houston, life went on in Concord.

Louie Cartier was a Concord High junior that year. He was sixteen. Other kids teased him about his dime-store clothes and chubby cheeks. "Chipmunk," they called him, the same nickname classmates had given Christa when she was a chubby grade schooler. Bullies called the boy "Chippie" for short and teased him about a football player who'd stuffed Chippie's mouth with toilet paper. "Hey, Chippie, guess who's waiting for you in the boys' room?" Cartier dropped out of school in the fall of 1985. That December, he reappeared in the hallway outside the principal's office, carrying a bottle of wine and a double-barreled shotgun.

"The gun didn't scare me," says Kevin Swope. "I saw him walking down the hall and thought, 'It's a prop for a play.'"

Cartier took two boys hostage. They were acquaintances he hoped to impress. Showing off the ammo belt draped over his chest and the live rounds in the gun, he smiled and said they were "fucking real." He herded the boys into a stairwell and offered them a slug of wine—Pink Catawba in a wicker-basket bottle. When they said no thanks, he smashed the bottle against a wall.

Minutes later, principal Charles Foley and football coach Don LeBrun hustled toward the disturbance. The coach had been Cartier's driver's-ed teacher. He said, "Louie, what are you doing?"

Cartier released one of his hostages but kept his shotgun aimed at the other. Soon policemen came running into the stairwell, shouting, "Put the gun down! Put it down!" One of them, Concord police sergeant John Clark, was a friend of the McAuliffes who had spoken to Christa's classes about his job, telling them how proud he was that he had never fired a weapon while on duty. Now he trained his own shotgun on Cartier.

Coach LeBrun offered to trade places with the hostage. "Why don't you let him go? How about I come take his place?"

Cartier waved the shotgun toward the coach. Officer Michael Russell, fearing that the boy would shoot LeBrun, fired his .38 revolver. Russell's bullet struck a cartridge in the youngster's ammo belt, knocking him back a few steps in a spray of sparks and smoke. "Why'd you hit me?" Cartier asked, firing back and grazing another policeman. Sergeant Clark fired twice, striking his target in the wrist and head. The head wound was fatal. Louie Cartier died the next day.

Principal Foley declared that he saw no reason to beef up security at Concord High. "What would you do, put a guard at all twenty-two entrances?" Foley asked reporters. According to the *Monitor*'s Ralph Jimenez, "We covered the story, but it didn't get much attention. It didn't resonate. School shootings weren't a thing yet. We all thought of it as an isolated incident, one of a kind."

Still, Christa was shaken. "When I heard, I was desperate," she said. "I felt like I needed to be at the school to help out." But there was nothing she could do after the fact, any more than there had been after young Todd Walker fell into the machinery of the Lunar Odyssey ride at Space Camp. There was nothing to do but keep going.

—◆—

Christa's months of training featured almost as many interviews as simulations. "The crew is unbelievable," she told *USA Today*. "Very professional, very concerned about how I am fitting in, trying to make me part of the team. Judy Resnik, for example, has a doctorate in electrical engineering. It's mind-boggling that she knows all these circuits and can figure all this out." Christa admired Scobee and Smith's test-pilot swagger, McNair and Onizuka's mastery of lasers and satellites, and her fellow payload specialist Jarvis's good humor and sense of wonder about the adventure they were sharing, but she was dazzled by Resnik, who seemed to have the rightest stuff of them all.

Born in Akron, Ohio, in 1949, Judy Resnik came from what was then called a "broken home." Her mother, Sarah, a legal secretary, and her father, Marvin, an optometrist who served as a cantor at Akron's Beth El Congregation, argued about everything. Sarah thought their daughter should learn to type and take dictation. Marvin doted on his *k'tanah*, his little one, showing her how radios worked by taking them apart and putting them back together. Even after she made astronaut she would phone Marvin and say, "Daddy, it's me, Little."

Both parents encouraged their daughter to play piano, which she soon did so well that her music teachers couldn't keep up. Her note-perfect recitals had the Resniks picturing their prodigy as a concert pianist. They were less enthralled by some of the boys she flirted with at a skating rink, some of them Gentiles. When Marvin bought her a pair of ice skates, Sarah threw them in the fire. After her parents divorced, Judy filled out her own custody papers. She moved in with her father and sued her mother for the bat mitzvah money Sarah was saving for her. Judy won the lawsuit. She was fourteen.

At Firestone High School, where she graduated first in her class, she took a typing class and was soon dashing off 125 words per minute. During a summer as a part-time typist at a law firm, "she made more money than anyone else because they were paid by the page," Marvin boasted.

Judy Resnik's grades and perfect SAT score brought scholarship offers from all over the country. She chose Carnegie Tech in

Pittsburgh over The Juilliard School in New York; she thought science was more practical than music. One of three women in a freshman class of ninety at Carnegie—later Carnegie Mellon University—she quickly switched majors from mathematics to electrical engineering. "College math got more and more abstract, and Judy was a practical person," recalls Michael Oldak, the college classmate she married in 1970, the year they graduated. As Judy Oldak she went directly into a doctoral program at the University of Maryland, skipping a master's degree. "We were both working for RCA in missile-to-surface radar," Oldak recalls. "RCA applied for several patents based on Judy's work. Pretty soon she got promoted. She got paid more than I did."

In 1974, the twenty-five-year-old PhD accepted a fellowship as a biomedical engineer at the National Institutes of Health. Her work there delighted her optometrist father, though Marvin found some of the details hard to follow. Judy's research had to do with the ways the eye translates light into electrical currents that travel to the brain to be reassembled as mental images. She would start by killing a frog and extracting the still-fresh retina—the layer of light-sensitive tissue that sends signals through the optic nerve to the brain—behind its eye. Shining a light on the tissue, she used a new technique to record the signals it transmitted. Her innovation was the basis of a scholarly paper she published in the *Journal of the Optical Society of America*, titled "A Novel Rapid Scanning Microspectrophotometer and Its Use in Measuring Rhodopsin Photoproduct Pathways and Kinetics in Frog Retinas."

Judy Oldak was one of the more prominent young scientists in her field, and she was miserable. University of Maryland professor Robert Newcomb, one of her mentors, would remember a brilliant but troubled woman. "She was going through a lot of turmoil. She felt she didn't have the freedom, as a woman, that she wanted," he said. "She felt constrained."

Mike Oldak wanted children. His wife didn't, at least not yet. According to a *Washington Post* profile that sized her up a decade later, "Many who knew her believe she immersed herself in her career to avoid motherhood." That was something no one ever said about astronauts and fatherhood.

"We wanted different things," says Oldak. "We got divorced in 1974, but stayed friends."

Professor Newcomb saw "a big change in her once she was divorced. A happiness." Resnik promptly reclaimed the name she was born with. Not her *maiden* name, a term she found laughable. "I'm no maiden!" As soon as her divorce was final she sent a greeting card to an old boyfriend: *I'm single.*

In 1977, she heard that NASA was looking for female astronauts. According to one of its press releases, the agency was now open to "women and other minorities." As she recalled years later, "I thought I'd give it a try. I had the scientific credentials to be selected. Besides that, I'm sort of a competitive person. If I want something, I *want* it."

She started with what she called "a bit of polite nosing around Washington," quizzing her fellow Ohioan John Glenn, now a US senator, and Apollo astronaut Michael Collins about what the agency was looking for. Collins said he wasn't sure. Describing one of his own psych tests, he recalled a NASA psychologist handing him a plain white sheet of paper. "What do you see?" the man asked. With no idea how to answer, Collins amused himself by saying, "Nineteen polar bears fornicating in a snowbank." The shrink must have liked him; Collins went on to serve as Commander John Young's pilot on Gemini 10 and joined Neil Armstrong and Buzz Aldrin on Apollo 11, the moon mission.

Resnik applied for the astronaut program. Months went by. The silence drove her to distraction. "Judy was used to choosing, not being chosen," a friend said. At last the Astronaut

Office announced that Judith A. Resnik was among two hundred women being considered for the astronaut class of 1978. While waiting to hear if she'd be chosen she took a job as an engineer at Xerox in El Segundo, California. She ran on the beach and shaped up with a high-protein diet. She lifted weights, which was practically unheard-of for women then. She took flying lessons, scoring two perfect 100s and a 98 on her pilot's-license exams. Astronaut Norm Thagard, a navy pilot who had flown 163 combat missions over Vietnam, called her "a natural flyer." Later, when she flew in the seat behind him in supersonic T-38s, Thagard let her take the controls. "And not just in flight. I let her take off and land us. That wasn't allowed, but that's how good a pilot she was."

Resnik was leaving her Redondo Beach apartment one morning when the phone rang. The caller was George Abbey, NASA's director of flight crew operations. "How's your morning going?" he asked.

"Fine."

Abbey knew every prospective astronaut candidate's heart leaped at the sound of his voice. He liked to keep candidates guessing. "How do you like working at Xerox?" Finally, he got to the point. "Would you like to do something different? Would you like to go fly?"

That summer, Resnik joined the first new astronaut class in a decade. It consisted of twenty-nine men and six women; they nicknamed themselves the "Thirty-Five New Guys," abbreviated TFNGs. That was an old military term that also stood for "The Fucking New Guys," which is what the new bunch represented to the dozens of veteran astronauts who had offices, desks, and lockers at JSC. The class of '78 included the first three Black AsCans, including Ron McNair, as well as the first Asian American and the first Buddhist. The final two categories described a single person, Hawaii's Ellison Onizuka, who described himself as a diversity double threat.

To the NASA veterans who set the tone in the Astronaut Office, the new arrivals were rookies and shavetails, minorities, and girls—until they flew.

They were told to report for training on Monday, July 10, 1978. Resnik got there early. "We came down to Houston," recalled Shannon Lucid, one of the six women in the class, "and she was already there, trying on spacesuits."

There was speculation about which of them would be the first female astronaut. Whoever it was would not be the first woman in space—that was cosmonaut Valentina Tereshkova, who made forty-eight orbits of Earth in *Vostok 6* in 1963. Yet American reporters seldom mentioned Tereshkova. To them, the US space program was the one that mattered. They would treat the first American woman in space as a national hero.

Other AsCans saw Resnik and Sally Ride as the frontrunners. Ride, twenty-seven, once a nationally ranked junior tennis player, had earned her PhD in physics from Stanford. One of the male TFNGs called her "the most in-your-face feminist" among them. As Hoot Gibson said, "I had pretty well pegged Sally and Judy as the first two." He recalled a day several of them reached JSC's Building 1 at the same time. "One of the boys grabbed the door and opened it for Sally, and Sally shoved him through and held the door herself." With the public affairs office eager to promote the expertise as well as the "femininity" of the first female astronaut, it was a testament to Resnik's and Ride's talents that the choice came down to an intensely private Jewish divorcee and an intensely private physicist who both came across as less traditionally feminine than their superiors may have liked. (Ride later married and divorced astronaut Steve Hawley; rumors that she was gay were not confirmed until her death in 2012.) After Ride got the call in 1982, Resnik was disappointed and relieved—disappointed that her dad wouldn't get to brag about his daughter's being the first female astronaut, and relieved because Ride would have to answer a thousand dumb questions that had nothing to do with the job.

During the run-up to Ride's ride, reporters kept asking her about bras and makeup in space. At one press conference, a writer from *Time* asked Ride if she cried when simulations went wrong. Turning to pilot Rick Hauck, she said, "Why doesn't anyone ever ask Rick these questions?" Commander Bob Crippen bailed them out with a joke: "The *commander* weeps," Crippen said.

Ride's 1983 flight drew almost a million spectators to Cape Canaveral, the largest crowd since the first shuttle missions. She returned to national and international acclaim, as well as to a bouquet of roses and carnations the public affairs office had waiting for her, which Ride refused to touch. She said she would accept flowers when the men got them, too.

Resnik applauded Ride's flawless performance on *Challenger*, then waited fourteen more months for her turn. By 1984, she had spent six years in training at JSC. She helped design and refine software for upcoming missions and spent countless hours learning to operate the shuttle's robot arm, which plucked satellites from the cargo bay and released them into orbit, a task she excelled at. That part of the job was a contribution to the space program that Resnik enjoyed. She couldn't say the same for her PR duties. As a representative of the astronaut corps who had yet to fly, she was often referred to in the press as an "astronaut hopeful." One feature story named her one of "the Glamournauts, six NASA lovelies." When a *Houston Post* writer asked if "the feminist movement" had opened doors for her, Resnik said, "No. I got here on my own merit." Still, the newsletter of NASA's Lewis Research Center, in Cleveland, described her as "the lady spaceman." Her duties included giving England's Princess Anne a tour of a shuttle simulator and presenting an award to actress Nichelle Nichols, who played Lieutenant Uhura on *Star Trek*.

She made her morning-TV debut on NBC's *Today* show. Sitting in a pink blouse beside *Today* anchor Tom Brokaw, her black curls freshly fluffed by a backstage stylist, Resnik fidgeted as Brokaw introduced her as "thirty-two years old, a native of Akron, holder of a doctorate in electrical engineering from the University of Maryland. She is single, she plays the piano, and she's a runner." If his intro made her sound like a contestant on *Love Connection*, at least he correctly called her an astronaut, a title she had earned as a 1979 graduate of the AsCan program.

Brokaw asked about life as a young woman at NASA. "Wasn't there a little bit of resentment or male chauvinism? That's a very male kind of fighter-pilot world you were entering."

She smiled. "Not at all. As a matter of fact, I think everybody leaned over backward to make sure we were treated as equals."

"Were you a tomboy when you were a kid?"

"No."

"What happens when you meet a man and you say, 'I'm an astronaut'? Does he say, 'Oh, you're too cute to be an astronaut. C'mon, little lady, you can't be an astronaut!'"

"I just say I'm an engineer."

Brokaw wasn't finished with the human-interest questions. "Do you think the time will come when there will be romance in outer space?"

She blinked. "Oh, gee, I really couldn't tell you about that. It's a career to us, and we treat it that way."

Her turn to fly came on *Discovery* in 1984. "Judith ('J. R.') Resnik may have been the most doggedly determined astronaut, male or female, ever to suit up," *Time* reported. In the weeks before the launch she deflected questions about her love life by claiming to be obsessed with TV star Tom Selleck. She hung a

Selleck poster in the women's locker room at JSC and drank coffee from a mug that read I'm Saving Myself for Tom Selleck. Her supposed crush on the hunky hero of *Magnum, P.I.* made for a simpler story than her real romantic entanglements. As friends knew, she was involved with an astronaut who was married. Another married admirer was her *Discovery* crewmate Mullane, who admitted being infatuated with Resnik. During training with their boisterous crew, the so-called Zoo Crew, he was "Tarzan" and she was "Jane." In his memoir, *Riding Rockets*, Mullane told of a night when the two of them mingled with members of Congress at a Washington, DC, reception along with astronaut Steve Hawley. "During a break," he wrote, "Steve whispered in my ear, 'Watch their eyes as they shake my hand.' I was confused by his comment, but only until the next senator passed. As the politician pumped Steve's hand, his head was turned and he was smiling directly at Judy. Steve was invisible. He could have greeted each of them with 'Kiss my ass, Senator,' and they would not have heard. They had come into the gravitational pull of Judy's beauty. . . . Judy handled it with her usual aplomb, equally gracious to the old lechers as well as the young ones."

Resnik looked forward to leaving reporters, senators, and 4.9 billion others behind on August 30, 1984, the day she became the second American woman in space.

To her surprise, she felt light-headed on the catwalk from the launch tower to the shuttle that morning. Here was something they hadn't rehearsed at JSC—walking across a catwalk 150 feet up. For a moment she felt like the troubled Akron teenager she had been sixteen years before, riding a roller coaster with a boy her mother disapproved of, trying not to throw up. Then as now, she steeled herself and got through it.

After that, spaceflight was easy. "Just like the simulators," she said. She taped a beefcake photo of Selleck near the million-dollar space toilet and made Marvin Resnik's year by holding a hand-lettered sign up to *Discovery*'s in-flight TV camera: HI DAD. During the shuttle's six days in orbit, Resnik worked so efficiently that she finished her duties ahead of time. Mission Control beamed up another to-do list, which she completed in time for reentry.

During free minutes in space, she took in the view from the flight deck. In real life the planet below was more colorful and alive than any simulation, with spring greens and fall browns in the temperate regions, blue oceans, white clouds, and the neon-green aurora borealis, the Northern Lights, swirling over the Arctic. One of her assignments was to assist crewmate Mike Coats as he deployed an IMAX camera to record footage of what they saw. To Resnik's dismay, several strands of her long, weightless hair got caught in the machinery, yanking her sideways. The camera's belt-driven magazine sucked up clumps of her hair. Mullane and Coats used scissors to free her, but the IMAX camera coughed and quit. Coats took it to the middeck for repairs.

Later, when the time came for Commander Hank Hartsfield to radio Mission Control, Resnik shot him a dark look. She didn't want to go down in history as the Glamournaut whose flowing locks jammed a multimillion-dollar space camera.

Hartsfield told Houston the camera "malfunctioned and is being repaired." Coats spent several hours picking strands of Resnik's hair from the camera, repaired it, and saved the footage they had shot, which provided many of the spectacular sequences in *The Dream Is Alive*, an IMAX film President Reagan called "the next best thing to flying in space."

The crew kept Resnik's secret from the public. Still, some of the TFNGs wondered if she would do the practical thing and cut her hair after such a close call. She didn't. "Judy was proud of her hair," says June Scobee. "She was proud of being very feminine *and* very professional." Resnik made no concessions to others' views of her even after a photo of the Zoo Crew in orbit drew complaints. "While we didn't intend it, the pose suggested a cheerleader's pyramid. Adding to the effect were Judy's legs. They dominated the photo," Mullane recalled, calling her legs "tan, perfectly proportioned, beautiful."

After the mission, a reporter asked Resnik how she maintained her lavish hairstyle in the vacuum of space. She said, "No comment."

By the time she met Christa, Judy Resnik was one of the most accomplished women in the history of aviation. "Christa looked up to her," June Scobee remembers. "They had a friendship, and in some ways Judy was her role model. Look at pictures from those days. Christa was trying to get her perm a little poofier. She wanted long, curly hair like Judy's that would float when she was in space."

8

NASA TRADITION CALLED FOR A PRESS CONFERENCE THIRTY days before each launch. With the holidays coming up, the public affairs office moved *Challenger*'s press conference forward by ten days. On Thursday the twelfth and Friday the thirteenth of December, the agency hosted more than a hundred reporters, the largest media contingent since Sally Ride's flight two years before. Flight director Randy Stone opened the proceedings with a slip of the tongue, calling their celebrity payload specialist "Christa McCoffee."

Stone ceded the floor to flight-training chief Frank Hughes, a veteran of the Apollo era. Asked if his simulation supervisors had put Christa McAuliffe through rigorous training for the mission, Hughes said, "No. In the early days it was a big deal, but now the shuttle is an old lady's ride."

"Have you had a chance to judge Christa's strengths and weaknesses?"

"Quite honestly, she's going to go up and do some pretty good experiments," he said, "but this is a lark. It's a big gee-whiz. She can look out the window all she wants. She's going to have the most wonderful time of her life."

Most of the writers and photographers were there to focus on Christa, but Resnik was the first member of the *Challenger* crew to face them. She looked as if she might be happier digging a ditch. As Hughes puts it, "Judy didn't like having to answer foolish questions." Sipping from a can of Diet Coke, she corrected a reporter who referred to her as a payload specialist. "*Mission* specialist," she said. When asked how training in simulators compared to spaceflight, she said the real thing was "just like you practice it, except you practice all the things that can go wrong, and usually nothing goes wrong."

Next came the first question about Christa. Would Christa McAuliffe "humanize the space experience" better than the career astronauts?

"I really don't know," Resnik said. "We're technical people. Part of Christa's job is to talk about it. Most of our job is not to talk about it, it's to do it."

"What about being a woman astronaut? Have you talked to Christa about that?"

"I don't think she needs any advice from me," said Christa's backstage mentor. After several more questions and curt replies, she left the room.

According to Hohler of the *Monitor*, "Scobee and Smith blew in like a fresh breeze." The commander and pilot both stood six foot one and weighed 175 pounds. Both had brown hair, blue eyes, and square jaws. Bantering with reporters who didn't know which of them was which, Scobee said, "All test pilots look alike."

When a writer repeated a cliché about the shuttle's being the most advanced vehicle in human history, a flying machine with more than two and a half million parts, he was ready with the punchline: "And all built by the lowest bidder." Yet Scobee wasn't content with amusing the press. His twenty-five-year career had led to this command. He felt a duty to represent his crew, the agency, the space program, and the United States as honestly and as well as he could. When a writer asked if there was a downside to spaceflight, Scobee took the question seriously. He said, "You return with a lot of memories. One of my consternations about coming back is that you have to tell everybody about it. Sometimes I'd just like to lock it up inside and keep it for myself."

Then the flyboys made way for the main attraction, who said she'd been sorry to hear that this would be the last press conference until they went to Florida for the launch. "I'm going to miss the interviews," Christa told the press corps, "because you people allow me to share what I'm doing. But everything's going to go real fast from here."

Asked about her space journal, she told the truth: she and NASA disagreed about who would hold the rights to what she wrote. "Nobody told me they were going to take it and publish it."

"So it's not going to be released?" a reporter asked.

"As far as I know, no. I'm not doing it for a lot of people. It's something I need to do for me. I want my kids to know what all of this has been like. Then, as a historian, I'd like to find a way to share it, maybe as a curriculum guide."

Next up, her fellow payload specialist Greg Jarvis described a fluid-dynamics experiment he would conduct in orbit, testing how liquid satellite fuels behave in orbit.

Then came mission specialists Onizuka, who would deploy a satellite that would study Halley's Comet, and McNair, the laser specialist who had become the second Black astronaut two years before. Along with the two women on the roster, they made Scobee's crew the most diverse NASA had ever assembled. The affable Onizuka joked that if anything went wrong with the shuttle's fifty-foot-long robot arm, he could climb out and fix it the way he'd shinnied up coconut trees as a boy in Hawaii.

McNair was not smiling. He was fuming at Scobee.

A PhD out of MIT, McNair had one of NASA's most remarkable life stories. He had grown up in Lake City, South Carolina, in a house that lacked indoor plumbing. His mother was a schoolteacher, his father a car mechanic. On stormy days the McNairs placed pots and pans around the house to catch rainwater dripping through the roof. "I come from a very upper-class family," he liked to say. "We just had a very low income." He and his two brothers obeyed the WHITE ONLY and COLORED ONLY signs on Lake City drinking fountains. "If we got 'uppity,' people in town made it their business to straighten us out," Ron's brother Carl recalled. Black kids sat in the balcony at the local movie house, where they were not allowed at the concession stand.

When Ron was nine, he tried to check two science books out of the Lake City library. "This library's not for coloreds," the librarian said. "If you don't leave right now, I'm calling the police." The boy said, "Okay. I'll wait." As Carl McNair remembered, their mother, Pearl, "comes down there prayin' all the way, 'Lordy Jesus, please don't let them put my child in jail.'"

Two white officers answered the librarian's call. "He's not causing trouble," one of them told her. "Why don't you let him have the books?" Finally, reluctantly, she stamped the books with a due date.

Pearl McNair nudged her son. "What do you say?"

"Thank you, ma'am," Ron told the librarian.

Several years later, as a member of a segregated Boy Scout troop, he saw Ku Klux Klansmen burn a cross in front of his troop leader's house. McNair spent summers picking crops in Florence County's cotton and tobacco fields. He went on to valedictorian honors at Lake City's all-black George Washington Carver High School, where he ran track and starred at middle linebacker for the football team. He passed up a football scholarship to Howard University to major in physics at North Carolina A&T State. After graduating with honors he landed a postgraduate slot at the Massachusetts Institute of Technology as part of a program that loaned professors to historically Black colleges while sending some of the best students from those colleges to MIT.

McNair struggled in Boston. He flunked his first exam. "Then he buckled down. He threw himself into his science," says Cheryl McNair, who married him in 1976. "And he amazed me. I'll never forget the first time Ron showed me a laser beam going through a tube. At that time most people had never seen one. It was bluish, then the beam turned red before it hit a metal sheet and burned through it." McNair earned his PhD in physics in 1977.

"People talk about born geniuses," said one of his college roommates, "but I always thought of Ron as a self-made genius. He got his through hard work."

Two years after earning his PhD, McNair applied to be an astronaut. As one of three Black AsCans in the class of 1978, he was in the running to be the first Black astronaut. Then, in 1982, he was driving on Interstate 45, south of Houston, with Cheryl in the passenger seat. "We stopped for a traffic jam," she recalls, "and a car hit us from behind. The other driver was doing sixty at least. *Boom!* I got knocked unconscious, but it was worse for Ron. His car door buckled into his body." McNair's fitness and martial-arts training may have saved his life. He had several broken ribs, internal injuries, and head trauma that left him with double vision. Doctors said he could expect to spend six weeks in the hospital. "But he was out in less than two weeks," Cheryl remembers. "That was Ron. He was in top shape, body and soul."

The accident knocked him out of the astronaut rotation. When the agency chose Guion Bluford to be the first African American astronaut in 1983, "Ron called me and wished me well," Bluford recalled. Would their roles have been reversed if not for the car crash? "We never knew. They never told us. It could have been a factor," says Cheryl McNair.

By 1984, McNair was fully recovered. "Stronger than ever," she says. Determined to prove he was fit, he stepped up his workouts. A compact five foot eight and 160 pounds, McNair was one of the fittest astronauts in NASA history, and he had a better party trick than Hoot Gibson's flaming drinks. A fifth-degree black belt in karate, McNair could break cement blocks with a single chop of his hand. *Scientific American* had published a study cowritten by McNair, "The Physics of Karate," explaining how a black belt like him could shatter concrete. "A well-executed karate strike delivers several kilowatts of power over several milliseconds," he and two coauthors noted. The others' academic credentials may have been more imposing, but it was McNair who starred in the article's most striking feature, a series of stop-action photos that showed the grimacing young physicist smashing his bare right hand through slabs of concrete. "How is it that the hand is not shattered by the force of the karate strike?" the authors asked. "Part of the answer lies in the fact that bone is much stronger than concrete. Consider how easy it would be to shatter a piece of concrete the size and shape of a bone. . . . The hand can withstand forces much larger because it is not a single piece of bone but a network of bones connected by viscoelastic tissue." The flexibility of bone and tissue allows a strong, fast-moving hand to "resist forty times more stress, or force per unit area, than concrete."

Mission specialist McNair's skills were not limited to laser physics and martial arts. As a saxophonist he sat in with touring jazz players when they passed through Houston and led an eighteen-piece swing band called Contra Band. He had stowed a miniature soprano sax on his first shuttle mission and played "What the World Needs Now Is Love." His crewmates recorded his performance, the first music video in the key of zero G, but they accidentally erased the tape. Two years later, McNair wanted to make sure nothing went wrong. His playlist included "Rendez-Vous," a tune he had written with composer Jean-Michel Jarre. Jarre booked a venue in Houston where concertgoers would be able to hear the tune Jarre called "Ron's song" as McNair played it during a simulcast from *Challenger*.

"It was going to be Ron's swan song," Cheryl McNair says. Once he recorded the first song in space, "He was going to retire from NASA to be a science professor at the University of South Carolina." The campus was an hour from Lake City, where his mother still lived. "He wanted his mother to see her son come home, an astronaut and a professor."

Scobee threw a wrench into McNair's plans by ruling that they would not be taking musical instruments into orbit. "We don't have room," he said. When McNair objected, Scobee heard him out. He wasn't convinced. "Tell you what," he said. "I'll let you take your saxophone if Judy can take her piano."

There was no point in arguing. A commander's word was law.

At the press conference, a *Dallas Morning News* reporter asked about McNair's saxophone solo. The mission specialist shook his head. It wasn't going to happen, he said.

"Why?"

"Let's say there's some objection." McNair wasn't going to criticize Scobee in public. "Someone in the chain of command objects to it."

"Is it too frivolous for a space mission?"

The question irked him. "It's not frivolous at all. It's anything but frivolous." When the reporter asked who had decided he couldn't take his horn, he said, "I'm not here to get anybody ticked off."

Another questioner changed the subject: "Christa has said nothing worries her about the flight, but the fear might hit her when she walks across the catwalk to get onboard. What could you tell her that might put her at ease?"

"I could tell her she's right," McNair said. "She'll walk out there and see those big boosters hanging there and the big tank and all that huge stuff smoking, and she's going to know it's for real."

While the conference wrapped up, the *Monitor*'s Bob Hohler found Scobee and Smith drinking coffee in a JSC hallway. Scobee explained that the saxophone issue was "no mystery. It was my decision. We don't have room." As for the payload specialist program that had added a pair of civilians to his crew, he admitted he'd had his doubts. "The idea really torqued me at first. We had people who had been waiting fifteen years to fly, and they lost their seat to somebody who walked in off the street." Yet he saw the publicity value to the program. Thanks to Christa, their mission was front-page news. He had told her she was the reason shuttle launches were back on TV. Scobee said he felt blessed to be working with the particular payload specialists he'd been assigned, hard workers he now considered full-fledged members of the Challenger Seven. Christa and Jarvis had turned him into a booster of the Space Flight Participant Program.

"All I ask is, don't fly me with a politician."

—◆—

By late December, the crew's training was wrapping up. It was agency policy to keep crews together in the weeks before they flew from Houston to Cape Canaveral for the last preparations for launch. "But Christa was smart. She figured out who to talk to," June Scobee says with a laugh. Christa told the mission's mother hen how much it would mean to her to spend a few days at home. June told her, "Let me have a word with the commander."

The week before Christmas, Christa was back in Concord, helping Steve and the kids decorate the house. She hung drapes in the master bedroom upstairs. She baked pies in a cinnamon-scented kitchen, wrapped presents, and handed out just-baked cookies to carolers. On December 22, the McAuliffes drove to Foxboro, Massachusetts, just south of Boston, to see the New England Patriots host the Cincinnati Bengals. Steve had taken Scott to several Patriots games, where they threw a football around the parking lot. Now Christa and Caroline joined them for the last Sunday of the 1985 regular season, a doubly special occasion because a victory would send the home team to the playoffs.

The thought of postseason football electrified New England fans, whose Patriots had never been to the Super Bowl. In fact the lowly "Patsies" had never won a single NFL playoff game. Tom Brady would be no help—their future quarterback was an eight-year-old peewee footballer in San Mateo, California. But after the home team took an early lead and went on to win, Patriots fans stormed the field. They tore down one of the goalposts and carried it through a stadium tunnel to the street, where the upright struck a twenty-thousand-volt power line. "Two fans were thrown through the air and landed on the opposite side of Route 1," newspapers reported. "Two others fell to the ground, and a fifth was thrown into the middle of the street." A Massachusetts state trooper arrived to find "bodies all over the place, and several hundred drunk people trying to tell state troopers what to do." Ambulances rushed the injured to a hospital, where they would all recover, while the party continued inside Sullivan Stadium. Nine-year-old Scott McAuliffe waited outside the home locker room, hoping to get a few autographs from his football heroes.

One of them was All-Pro tackle Brian Holloway. At six foot seven and 285 pounds, Holloway dwarfed most of the crowd around him. "We'd just won the biggest game of the season," he remembers. "Everybody's going crazy, and we're surrounded by Boston royalty—the Sullivan family, who owned the team, some of the Kennedys, Roger Clemens, Bobby Orr. In the midst of it all I saw a smiling face. I couldn't place her at first, but I was kind of a science geek. I'd seen Christa's TV appearances. I said, 'Hey, aren't you the Teacher in Space?'"

To Scott's amazement, Holloway asked for his mother's autograph. "But I didn't have any paper!" Holloway says. "I ran to Raymond Berry's desk, which is *not done*—you don't go bursting into the head coach's office—and grabbed some stationery and ran back to Christa."

She signed her name *S. Christa McAuliffe* on New England Patriots Football Club letterhead, adding *Brian—Reach for the stars! I'll be there!*

Holloway introduced his teammates to their celebrity guest. As one Patriot joked, "She was stealing the spotlight. She was a bigger celebrity than those of us who played in the game."

Three days later, the McAuliffes attended Christmas Mass together. Steve gave Christa a pair of gold earrings shaped like apples. She would wear them under her helmet on launch day. She spent New Year's Eve—"First Night," New Englanders call it—judging an ice-sculpture contest on New Hampshire's statehouse lawn. It was the warmest First Night in years, almost forty degrees—sweater weather in Concord. Church bells rang at midnight, and it was 1986.

———◆———

Returning to Houston for the crew's final weeks of training, Christa bet Scobee a beer that the Patriots would beat his Super Bowl choice, the Los Angeles Raiders, in the playoffs. New England won. A week later, with linemen John Hannah and Holloway bowling over Dolphins defenders, the Patriots thumped Miami in the American Football Conference championship game. They were going to the Super Bowl.

That Sunday, January 12, was the day the shuttle *Columbia* finally carried US Representative Bill Nelson aloft to look for angels, but only after seven delays in less than a month. As various glitches canceled one liftoff after another, Tom Brokaw had pronounced *Columbia* "zero for five." CBS anchor Dan Rather said, "The launch has been postponed so often that it's now known as Mission Impossible." *Columbia* landed safely six days later, but its scrubs had a ripple effect. The Teacher in Space mission, initially scheduled for January 22, was pushed to the twenty-third, twenty-fourth, and then the twenty-fifth. Each postponement added pressure to a space agency eager to prove its value to Representative Nelson, Senator Garn, and the agency's other Congressional overseers, as well as the press and public. As NASA Administrator James Beggs had told *Aviation Week and Space Technology* in the spring of 1985, "The next eighteen months are very critical. If we are going to prove our mettle and demonstrate our capability, we have got to fly." Beggs admitted privately that the agency could effectively manage only "around a dozen" shuttle flights per year. There were fourteen more, including the Teacher in Space flight, lined up for the rest of 1986, with nineteen on the docket for 1987. Such a schedule kept the pressure mounting on everyone from Beggs down to shuttle-maintenance crews at Cape Canaveral. To keep up, some KSC technicians worked up to eighty hours a week for seven or eight weeks in a row.

Within hours of *Columbia*'s fiery reentry on January 18, NASA executive Jim Harrington dispatched a supersonic T-38 to Edwards Air Force Base in the Mojave Desert. As the shuttle program's flow director, Harrington was responsible for assuring that all four of the so-called orbiters in the shuttle fleet—*Columbia*, *Challenger*, *Discovery*, and *Atlantis*—had the hardware they needed to fly efficiently and safely. But with so many missions so close together, the government contractors that manufactured orbiter components fell behind schedule. NASA answered the need by transplanting pieces of one shuttle to another. The agency's own term for this process was "cannibalization." After Commander Hoot Gibson glided *Columbia* to a landing at Edwards, workers removed forty-five key components, including a nose-wheel steering box, an air sensor for the crew cabin, and one of the shuttle's five onboard computers, and loaded them into the T-38. The needle-nose jets had a small freight compartment called a travel pod, but in cases like this the pod was too small to hold all of the components. Some had to be packed into boxes that were buckled into the seat behind the pilot for the cross-country flight to Florida. The T-38 flew the parts to Cape Canaveral, where a van delivered them to the Vehicle Assembly Building (VAB), NASA's garage.

Flow director Harrington heard from Mike Smith that night. *Challenger*'s pilot made a habit of studying the weather charts NASA's meteorologists churned out, and he didn't like what he saw. According to Malcolm McConnell, who covered the space program for *Reader's Digest*, "Smith was so alarmed about the weather" that he sneaked out of the astronaut quarters at JSC, where he and the other six were quarantined in the days before they flew to Cape Canaveral, "so that he could use the telephone in privacy." Smith told Harrington, "The weather looks terrible for next week. You've got to do everything you possibly can to install those spares, finish the work on *Challenger*, and get us launched by Saturday, or Sunday at the latest." He had already warned his family and friends traveling to Florida that they should expect to spend "up to two extra days in case of bad weather or shuttle problems." If *Challenger* didn't launch by Sunday the twenty-sixth, he thought it might be February before the weather cleared.

With less than a week remaining, technicians plugged pieces of *Columbia* into *Challenger* in the Vehicle Assembly Building, one of the largest buildings in the world. The VAB covered eight acres and stood 525 feet tall. Low clouds sometimes formed around its upper reaches. Inside, KSC workers rode cherry pickers to inspect the 122-foot shuttle's engines and interior as well as its seals, hatches, and windows, and the more than twenty thousand six-inch silica tiles that would protect it from the heat of reentry. Those black tiles, made from the purest quartz sand, fitted and glued by hand onto the shuttle's fuselage, were so resistant to heat that if you held one by the edge when the center was red hot, you wouldn't feel a thing.

When *Challenger* was ready to roll, the VAB's doors—each half again as tall as the Statue of Liberty—began to open, a process that took forty-five minutes. At last the stack emerged: *Challenger* standing in its vertical launch position, mated to its huge rust-colored fuel tank and silo-shaped rocket boosters, all riding the crawler, a specialized bulldozer mounted on tank treads. The largest land vehicle on earth, NASA's crawler delivered the billion-dollar stack to the launchpad, a distance of four miles, without jarring it. Topping out at two miles per hour, it took two and a half hours to make the trip.

Now came another delay. With launch day approaching and overworked ground crews behind schedule, *Challenger*'s launch was put off until the twenty-sixth, the day the Patriots would face the Chicago Bears in Super Bowl XX.

On Thursday the twenty-third, astronauts Scobee, Smith, Resnik, Onizuka, and McNair climbed into three T-38s and jetted out of Houston, darting over the Gulf of Mexico toward Florida. Payload specialists McAuliffe and Jarvis took a slower route. They joined Cheryl McNair, Lorna Onizuka, June Scobee, and Jane Smith in the "wives' plane," a Gulfstream known as "NASA Two," which met the sleek T-38s at Cape Canaveral. "A fun flight," June Scobee calls it. "After so much waiting we were all excited to get going. We were teasing and joking, Christa as much as anyone, but she was working, too. Here we were enjoying some of the last hours before the flight, and she spent them at a table in the plane, writing to Steve and their kids, writing letters of recommendation for students. I loved her already—everybody loved Christa. But I admired her for that."

Reader's Digest's McConnell, one of the finalists in the Journalist in Space program, waited on the runway at KSC. He watched as the three T-38s "taxied in an arrowhead pattern toward the waiting press. Behind them NASA Two, the Gulfstream executive jet, set down. . . . Commander Dick Scobee piloted the lead plane, with mission specialist Ellison Onizuka in the back seat. Judy Resnik flew behind shuttle pilot Mike Smith. The photographers' motor drives whined as Dr. Resnik removed her helmet and her dark hair fell to the shoulders of her blue flight suit. From the third jet, mission specialist Ron McNair climbed down with athletic grace." To McConnell, this was "about as heterogeneous a group of Americans as could be assembled." It was also a happier group than the one that had faced the press six weeks earlier in Houston. Even McNair was jazzed to be back at the Cape.

It was almost dusk. After a public affairs officer set up a microphone, Scobee spoke first. "It's a pleasure to be at the Cape," the commander said, standing with his hands on his hips, "to participate in something the Cape does better than anybody in the world, and that's launching space vehicles." The sky was clear, the temperature fifty-eight. "We expect weather like this on Sunday when we launch."

Next he introduced his second in command, calling Smith "one of the best flyers in the world." Smith noted that four of his crewmates had been to space before, while he and the payload specialists were rookies. "We're looking forward to getting the secret handshake," he joked. Next up, Resnik and Onizuka said how glad they were to be at the Cape after four months of training in Houston.

Then came McNair. "Ron didn't appreciate being told he couldn't take his saxophone, but he was going to leave NASA with no regrets," Cheryl McNair recalls. On the tarmac at KSC, McNair said, "I'd like to echo the feelings of the other crew members. And I'd like to introduce you to the person perhaps that you came to see. That's Christa McAuliffe, our payload specialist Teacher in Space."

Stepping to the mic, Christa said, "I am *so* excited to be here, and I hope everybody tunes in to watch the teacher teaching from space!" She introduced Jarvis, who often joked that being the *other* payload specialist on her flight made him a human anticlimax. After getting bumped off one shuttle mission by Senator Garn and another by Representative Nelson, he said, "It's a great pleasure, finally, to get this far."

The crew waved to family, friends, and reporters. They were driven in darkness to KSC's Operations and Checkout Building, a sprawling steel-and-glass box, and shown to their dorm-style rooms on the third floor, where they would be quarantined until the launch. The astronauts' quarters in the O&C Building hadn't been redecorated since Neil Armstrong and Buzz Aldrin bunked there twenty years before. There were a dozen bedrooms, a conference room, six unisex bathrooms, a kitchen and dining area, and a gym with weights and stationary bikes. Their windows looked out over palm trees and dark swampland and, in the distance, the launch tower where the stack stood: the orca-shaped shuttle with its rocket boosters and 180-foot-tall fuel tank, lit up by floodlights.

Greg Jarvis remembered the first time he had seen a space-shuttle stack up close. Looking up at the leviathan tethered to the tower, he took a moment to catch his breath. "The only thing that's small around here," he said, "is us."

9

T HE ORBITER WAS SMALLER THAN THE RUST-COLORED FUEL
tank it would ride piggyback into orbit. The external tank (ET),
which would be filled with liquid hydrogen and liquid oxygen on
the morning of the launch, served as an enormous thermos bottle
for the supercooled fuel it contained. The ET had been painted
white for the first two shuttle missions. Then an audit showed
that the latex paint required to cover it weighed six hundred
pounds. NASA's rule of thumb was that it cost ten thousand dollars to send a pound of anything into orbit, whether that thing
was a freeze-dried dinner, a layer of paint, or part of a satellite or
human being. That made the paint on the tank worth six million
dollars. After the audit, ETs went unpainted.

The four orbiters in the shuttle fleet—*Columbia, Discovery,
Challenger*, and *Atlantis*—were nearly identical. Each was 122

feet long with a wingspan of 78 feet, about the size of a Boeing 727. Each featured a small flight deck—the shuttle's cockpit—with panoramic windows. A larger middeck below the flight deck functioned as the crew's living quarters. The much larger cargo bay behind the flight deck and middeck measured nine hundred square feet, enough to hold a Greyhound bus, or the satellites that were *Challenger*'s primary payload. One of the two satellites would help NASA communications by linking other satellites orbiting the Earth. The other, SPARTAN/Halley (SPARTAN for Shuttle-Pointed Autonomous Research Tool for Astronomy), was a 2,250-pound box the shape of a window air conditioner designed to study Halley's Comet in ultraviolet light, one of McNair's specialties, for the first time. Once *Challenger* escaped Earth's gravity, Resnik would use its fifty-foot robot arm to pluck the satellites from the cargo bay and drop them into orbit.

Built by Rockwell International in Palmdale, California, the orbiters were often described as the most complex machines ever built. The ET, their giant fuel tank, and twin rocket boosters fueled their climb through the atmosphere and then fell away as the orbiters went on to circle the planet at a friction-free seventeen thousand miles per hour, often upside-down. (In weightless space, upside-down and right side up are the same.) On the downhill, after reentry, a space shuttle functioned as a glider, making it tricky to handle for all but the best flyers. One commander said maneuvering the shuttle to a safe landing was like "flying a brick." *Columbia* and *Challenger*, the first operational orbiters, were heavier than *Discovery* and *Atlantis*, which were built with newer, lighter alloys. When fully fueled and loaded with its satellite cargo, *Challenger* would be the heaviest shuttle on record, weighing precisely 4,529,122 pounds. It would be so heavy that the three main engines at the orbiter's tail could supply only 20 percent of the thrust required to lift it into space. "The other eighty percent came from the rocket boosters," says engineer Lee Solid, who helped build the shuttles' engines. "It takes a lot of force to get something that massive into orbit." It would take the combined power of the boosters and main engines to lift *Challenger* into space; once there, the crew would rely on the small onboard rockets of the orbital maneuvering system (OMS) to make course corrections.

Like evolution, astronautical engineering proceeds by building on what came before. "Engineers hardly ever start from scratch," NASA historian Bill Barry explains. "They take what's been done before and try to make it better." *Challenger* was the result of refinements on Mercury, Gemini, and Apollo engineering as well as a decade's worth of shuttle prototypes and test flights. The same could be said of the agency itself in the 1970s and '80s. The shuttle program was NASA's attempt to engineer its future after the moon landing of July 20, 1969.

—◆◆—

The moon shot was America's goal and great triumph of the 1960s. With six months to spare, Apollo 11 made good on Kennedy's pledge to put a man on the moon "before this decade is out." But Armstrong's "one small step" was also the ultimate step of the space race. After Armstrong, Aldrin, and Collins returned to Earth, what was the point of the space program?

There were billions of dollars and thousands of jobs at stake. NASA's budget had reached $5.9 billion at the height of the space race in 1966, accounting for 4.4 percent of federal spending. By 1972, its budget was 1.5 percent of federal spending and falling. Over the same six years, NASA's workforce fell from 396,000 employees to 160,000. (It would be 17,000 in 2021.) Nineteen seventy-two was the year the last of the dozen men to walk on the moon, Apollo 17 astronauts Gene Cernan and Harrison Schmitt,

Steve and Christa McAuliffe at home with son, Scott, and daughter, Caroline.

(Associated Press)

Vice President Bush named Christa NASA's "Teacher in Space" at the White House on June 19, 1985.

(George H. W. Bush Presidential Library and Museum)

She met Commander Dick Scobee (right), pilot Mike Smith, and mission specialist Ron McNair on her first visit to the Astronaut Office in Houston. (NASA)

Astronaut McNair was a black belt who smashed concrete blocks with karate chops.

(*American Journal of Physics* with permission of the American Association of Physics Teachers)

After suffering through her first rides on the "Vomit Comet,"
Christa got accustomed to weightlessness. (NASA)

She joined
the crew for
briefings during
training. (NASA)

Judith Resnik floated above Commander Hank Hartsfield (center), fellow mission specialist Mike Mullane (middle left), and the rest of the *Discovery* crew. (NASA)

Five of "the Female New Guys"—from left, Sally Ride, Resnik, Anna Fisher, Kathryn Sullivan, and Rhea Seddon—vied to be the first female astronaut. (NASA)

Astronaut Resnik sent her father a message from the space shuttle *Discovery* in 1984. (NASA)

After four months of drills, Christa couldn't wait to fly—
and then get back to her classroom. (NASA)

The official mission patch featured an apple for the Teacher in Space. (NASA)

During training, Scobee let Christa take the controls of a supersonic jet—that's Houston over her right shoulder. (NASA)

A flight-deck simulation tested (left to right) pilot Smith, mission specialists Ellison Onizuka and Resnik, and Commander Scobee. (NASA)

Scott McAuliffe was amazed when the Patriots' Brian Holloway asked for his mother's autograph.
(BrianHolloway.com)

A holiday party found Christa and Barbara Morgan in front of Jane Smith, Cheryl McNair, June Scobee, and Lorna Onizuka.
(Courtesy of June Scobee Rodgers)

Scobee told Christa about the sights she'd see from orbit, including the Northern Lights over the midwestern United States. (NASA)

Challenger rode the Shuttle Carrier Aircraft, a customized 747, to Cape Canaveral. (NASA)

The official crew photo found Smith, Scobee, and McNair in the front row with Onizuka, McAuliffe, Greg Jarvis, and Resnik behind them. (NASA)

The crew let Christa think she wouldn't be in their photo, then dressed up as schoolchildren to surprise her. Morgan, Christa's Teacher in Space backup (top right), subbed for Jarvis. (Courtesy of June Scobee Rodgers)

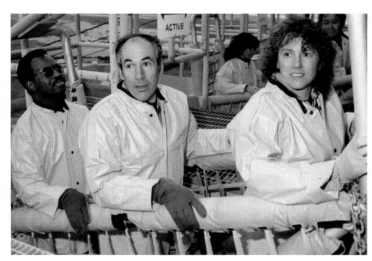

McNair and Jarvis followed Christa in a launch-tower drill at Cape Canaveral. (NASA)

NASA moved its shuttle stacks, with their giant fuel tanks and million-pound rocket boosters, from the Vehicle Assembly Building through doors taller than the Statue of Liberty. (NASA)

NASA's crawler, the biggest land vehicle on Earth, delivered *Challenger* to the launch pad at a top speed of two miles per hour. (NASA)

Christa sat beside Jarvis on the middeck with McNair by the hatch behind them. (NASA)

The crew met for launch-morning breakfast with a cake iced with the mission patch. (NASA)

Icicles hung from the launch tower on the morning of January 28. (NASA)

Scobee and Resnik led the way to the Astrovan. (NASA)

At liftoff a puff
of dark smoke
showed a leak in
the right-hand
rocket booster.
(NASA)

The leak led to a plume of flame above the booster's exhaust. (NASA)

The rocket boosters kept flying after the stack broke apart. (NASA)

President Reagan
watched replays of the
Challenger explosion
with his aides in the
White House.
(Ronald Reagan Presidential
Library)

At the *Challenger* memorial
service in Houston, First
Lady Nancy Reagan took
June Scobee's hand.
(Ronald Reagan Presidential Library)

Nobel Prize–winning physicist Richard Feynman used a glass of ice water to
demonstrate how cold could stiffen O-rings. (Associated Press)

Backed by family members Cheryl McNair, Jane Smith Wolcott, Chuck Resnik, and Lorna Onizuka, Dr. June Scobee Rodgers built a network of Challenger Learning Centers to "continue the mission." (Courtesy of June Scobee Rodgers)

NASA workers entombed the shuttle's debris in nuclear-missile silos at Cape Canaveral. (NASA)

More than five million students have learned STEM skills at Challenger Centers in the United States, United Kingdom, Canada, and South Korea. (Courtesy of Challenger Center)

S. CHRISTA McAULIFFE
SEPTEMBER 2, 1948 — JANUARY 28, 1986
WIFE. MOTHER. TEACHER
PIONEER WOMAN
CREW MEMBER, SPACE SHUTTLE CHALLENGER
AMERICA'S FIRST ORDINARY CITIZEN TO VENTURE
TOWARD SPACE
SHE HELPED PEOPLE. SHE LAUGHED. SHE LOVED AND IS LOVED. SHE APPRECIATED THE WORLD'S
NATURAL BEAUTY. SHE WAS CURIOUS AND SOUGHT TO LEARN WHO WE ARE AND WHAT
THE UNIVERSE IS ABOUT. SHE RELIED ON HER OWN JUDGMENT AND MORAL COURAGE
TO DO RIGHT. SHE CARED ABOUT THE SUFFERING OF HER FELLOW MAN. SHE TRIED
TO PROTECT OUR SPACESHIP EARTH. SHE TAUGHT HER CHILDREN TO DO THE SAME.

Visitors leave school and space mementoes at Christa's grave in Concord. (Tabitha Hawk)

left a stainless-steel plaque in the gray dust of the lunar surface. HERE MAN COMPLETED HIS FIRST EXPLORATIONS OF THE MOON, DECEMBER 1972 A.D. Nobody has set foot on the moon since.

With the Vietnam War escalating, Richard Nixon's administration wanted to pare back NASA's budget. Nixon's Scientific Advisory Committee called for a "space effort about half the magnitude of the present level." There was talk of a mission to Mars, but the red planet is three hundred times as far away as the moon. Getting there would call for another decade-long commitment that might cost five times as much as the budget-busting Apollo program. Instead, Nixon's advisors recommended a less expensive "reusable Space Transportation System" that would collaborate with other nations and US aerospace firms and eventually pay for itself.

As one astronaut put it, "NASA sold Congress on the premise that the space shuttle would make space cheap." Rockwell landed the contract to build America's "new reusable spaceplane" by submitting the lowest bid: five orbiters at $1.5 billion each, a fleet Jimmy Carter trimmed to four after he became president in 1977. The Carter administration slashed NASA's budget while diversifying the astronaut ranks with the TFNG class of '78.

The agency's budget rose after Ronald Reagan ousted Carter in the 1980 election. "Reagan's people were generally in favor of the space program," says space-program historian John Logsdon. "I think Reagan was captured by the romance of it." The new president liked the thought of putting his own stamp on the final frontier. His advisors liked the idea, too, if only to make space a "Reagan cause" in case John Glenn chose to run for president in 1984. At the same time, Reagan had run as a budget cutter. He subscribed to the so-called Nixon Doctrine that treated NASA as one of many federal agencies competing for funds. "Whatever new initiatives might be undertaken in space," Logsdon says, "they would have to be carried out without a significant increase in the budget." With the shuttle program up and running by 1981, the "space truck" looked as if it might help Reagan get reelected.

The president flew to California to welcome *Columbia* back to Earth on the Fourth of July 1982, putting paid to the shuttle program's first official flight. Logsdon recalls seeing President Reagan and First Lady Nancy Reagan "on a flag-bedecked platform" at Edwards Air Force Base along with astronaut Bob Crippen, "searching the sky for a glimpse of the orbiter." The president wondered where the shuttle was at that moment. Crippen checked his watch and said, "Over Hawaii."

A pair of sonic booms heralded its approach. Four hundred thousand spectators, sitting in and on their parked cars, trucks and RVs as if at a drive-in movie, watched *Columbia* glide toward Edwards's 2.8-mile runway under the command of Apollo program veteran Ken Mattingly (played by Gary Sinise in the movie *Apollo 13*), with Hank Hartsfield in the pilot's seat. Hartsfield called out altitude readings on the way down: *"Fifty feet . . . ten feet, one foot, half a foot."* The shuttle's three pairs of double wheels touched down so lightly that Mattingly thought they were still in the air. He called it the smoothest landing of his life—with the president there to see it. Feeling like the king of the runway, he yanked off his helmet and stood up too fast, whacking his head on a control panel. The reeling, triumphant commander had blood running down his face.

Hartsfield said, "That was very graceful."

Mattingly would recall it as "terrible. *Oh*, did I have a headache."

Commander and pilot got their earth legs back in time to join the Reagans on a platform near the runway. The president seemed impressed and even moved by the shuttle's return. "I think all of us feel a real swelling of pride in our chests," he told a cheering crowd that included singing cowboy Roy Rogers and thirty-five-year-old movie director Steven Spielberg. "Today we celebrate the two hundred and sixth anniversary of our independence. The space program in general and the shuttle program in particular have gone a long way to help our country recapture its spirit of vitality and confidence. The pioneer spirit still flourishes in America. . . . Beginning with the next flight, *Columbia* and her sister ships will be fully operational, ready to provide economical and routine access to space for scientific exploration, commercial ventures, and tasks related to national security." Better yet, a second orbiter was ready for action. "Way out there at the end of the runway," Reagan said, "the space shuttle *Challenger* is about to start on the first leg of a journey that will eventually put it into space. . . . It's headed for Florida now, and I believe they're ready to take off."

Challenger, fresh out of Rockwell's assembly plant in nearby Palmdale, sat on top of NASA's Shuttle Carrier Aircraft, a modified 747 that would fly the shuttle piggyback to Kennedy Space Center for its maiden flight the following year. The SCA was a unique vehicle in its own right. To balance the hundred-ton shuttle mounted toward its tail, its first-class section had been stripped of seats and filled with dumpsters full of gravel for ballast.

"*Challenger*," Reagan announced, "you are free to take off."

Logsdon described the next minutes: "As the huge airplane carrying *Challenger* lifted off and flew by the president, dipping a wing in salute, a band played 'God Bless America.' Reagan remarked, 'This has got to beat firecrackers.'" Turning to NASA Administrator James Beggs, Reagan called his day at Edwards "the most fun I've had since I got this job."

—◆—

Budget-conscious executives like Beggs hoped the shuttle program might come close to paying for itself by carrying satellites into orbit for American and foreign corporations. In 1983, a NASA brochure distributed at aerospace conventions called the shuttle "the most useful and versatile space transporter ever built." The glossy brochure carried one of the agency's marketing slogans: *"In all the world, you won't find the Shuttle's equal . . . and you can't get a better price."*

The earliest shuttle missions carried bare-bones crews of two: commander and pilot. Later flights added mission specialists to deploy satellites and perform in-flight experiments. Mission specialist Sally Ride joined a five-person crew on the seventh shuttle flight, in 1983. Mission specialist Guion Bluford broke the astronauts' color line as part of a five-man crew four months later. By 1985, the typical shuttle complement had grown to seven: commander, pilot, three mission specialists, and two payload specialists. The last category was created largely to reward influential politicians and billion-dollar contractors by giving them the priceless experience of spaceflight. And why stop there?

By the mid-eighties, Beggs was looking for a journalist to join the Space Flight Participant Program, someone who would give NASA "if not good publicity, at least a lot of publicity." Public relations had helped fuel the agency since *Life* magazine portrayed the Mercury astronauts and their wives as heroes and heroines of America's space race. Half a dozen presidents also played key roles. Two and a half years after his pledge to reach the moon by 1970, Kennedy told Texas congressman Albert Thomas that he himself was NASA's biggest booster. "The space program needs more identification," Kennedy said. "I plan on going to Cape Canaveral to give it some." Later that day, in Dallas, he was shot and killed. Within days, Jacqueline Kennedy asked Lyndon Johnson to commemorate her husband's devotion to the space program by renaming Cape Canaveral in his honor. The new president did just that in a subdued Thanksgiving address to the nation a week after the assassination. "The NASA Launch Operation Center in Florida shall hereafter be known as the John F. Kennedy Space Center," Johnson declared. Cape Canaveral was rechristened Cape Kennedy, a name that lasted for almost exactly a decade—until 1973, when the Florida legislature restored the low-lying cape's traditional name, which is Spanish for "bed of reeds." That same year, the Manned Spacecraft Center outside Houston was renamed in Johnson's honor.

Americans were losing interest in the space program. By the time Christa joined the *Challenger* crew, a young Jerry Seinfeld was asking Johnny Carson's audience, "The space shuttle . . . is it back?" Seinfeld wasn't sure. "Right now," he said, "people don't even follow the space shuttle! They should make it exciting. Maybe they should send up some guy that *doesn't* want to go. Everybody would watch that: dragging some guy down the hall in his spacesuit, he's holding onto doorjambs"—Seinfeld mimed mushing his terrified face against a space-shuttle window. "You'd see his face in the glass on liftoff . . ."

Christa was helping NASA make spaceflight exciting again. All three TV networks ran human-interest stories about her. CNN, the all-news cable channel founded by Ted Turner in 1980, announced that it would televise the Teacher in Space flight. PBS would carry her space lessons live to thousands of classrooms.

The day after she settled into her dorm room in the O&C Building, Christa joined the Scobees and other *Challenger* families at a NASA-owned beach house east of KSC, a private refuge at the end of a sandy dirt road, where generations of astronauts had dined and danced in the days and nights before launches. A wooden cottage with a brick fireplace and a deck facing the Atlantic, the beach house was so close to the launchpads that it had to be evacuated before shuttle launches.

"We took a walk on the beach with the Corrigans, Christa's mother and father, and Judy Resnik's dad and stepmother," June Scobee recalls. "Betty Resnik wanted to make sure I understood

Judy's achievements as an engineer and astronaut, even if Christa was getting all the attention. Just then, Judy came running toward us."

Grace Corrigan remembered the same moment. "Among my favorite memories is of Judy racing across the sand," she wrote. "She looked like such a little girl as she hurled herself into her father's arms."

Later, over fried chicken and deviled eggs at a picnic table on the deck, Dick Scobee talked about the marvels they would see from space. "You'll see things only a few people have ever seen. When you're orbiting at seventeen thousand miles an hour, the sun rises and sets every three hours. You can look down and see the flashes of lightning in thunderstorms. You can look right into the eye of a hurricane." The view would be just as brilliant at night. When the orbiter was on the night side of Earth, with no atmosphere or terrestrial lights polluting the view, the night sky would be dotted with five times as many stars as earthbound viewers saw—ten thousand stars, compared to two thousand visible from earth, ten thousand stars that do not twinkle. The twinkling of stars is an atmospheric effect; as seen from space they are hard points of light, the sky so thick with them that the familiar constellations seen from Earth are almost impossible to pick out. "And if you look down at the right time, you might see a shooting star." Rocks ranging from pebble to snowball size flew through space for millions of years, only to catch fire as they crashed into Earth's atmosphere just below the shuttle's orbit.

Scobee left out some details Christa's high schoolers would have liked to hear. Among the least publicized aspects of any mission was the engineering of human waste. Urine, for instance, is heavy. Rather than stow gallons of it on weeklong missions, shuttle crews opened a valve and released it into space, creating an unprecedented sight: a crew's urine spewing from a nozzle connected to the toilet and crystallizing instantly in the vacuum. Gordon Cooper, one of the original Mercury astronauts, described a spacecraft's urine trail as "a beautiful gold color," one of the more striking sights in space.

Resnik had uncomfortable memories of *Discovery*'s waste collection system. During her 1984 flight a heating element on its external nozzle failed. As a result, "This blob of urine froze on the side of the vehicle," wrote Mullane, who dubbed the blob "the urine-sicle." Mission Control fretted that during reentry it might "break off and fly back, hit the tail, gouge out the heat tiles, and the Shuttle would crash. I had thought of my life being threatened in many ways," he recalled, "but never by a block of frozen urine." According to Commander Hartsfield, Mission Control "came up with the idea of breaking it off" with the mechanical arm that placed satellites in orbit. Resnik helped the commander guide the fifty-foot robot arm, whose software she had helped design. "Judy handled the communications and kept me honest, double-checking procedures," while Hartsfield chipped away at the frozen blob. After a few moments, the urine-sicle came loose and drifted away, possibly saving all their lives. As Mullane remarked, "We'd better hope any aliens that are in orbit go by the old Boy Scout adage, 'Don't eat yellow snow.'"

At the beach house, Resnik didn't talk much about her six days and ninety-seven orbits of Earth on *Discovery*. It wasn't that she objected to pee talk. She was the potty mouth of the female astronauts, if not of all of them. She let others reminisce without mentioning the inelegant sequel to the urine-sicle story: with the space toilet out of commission, the men could direct their weightless urine streams into bags. The only woman onboard had to bunch up towels, pee into them, and store them in a trash bag for the three days until they landed.

Scobee described reentry as "an amazing light show, with the

fires of hell burning outside the window." He said he'd be preoc-
cupied flying the orbiter during most of their week in space, but
hoped to keep tabs on Christa's lessons and the experiments she
and the others conducted. As Crippen's second in command two
years earlier, he said, he had been captivated by an onboard experiment informally called "Bees at Zero Gs." The 3,400 honeybees
in a colony stowed on the middeck "were disoriented at first,"
Scobee remembered. With no air for their wings to beat against,
the bees couldn't fly. "They gained control by linking together in
a chain and crawling over each other." By the time he and Crippen brought the shuttle home, the bees had made honeycomb in
their locker on the middeck.

At the beach house, Christa said she couldn't wait to get off
the launchpad. At the same time, she found herself thinking more
about the hazards of spaceflight. She confided her worries to June
Scobee. "Does it scare you?"

"Who *wouldn't* be scared?" June asked. She had often heard
her husband say that any sane person would feel "a healthy fear"
of trusting life and limb to a vehicle that had two and a half million parts. As Christa knew from her background reading, more
than seven hundred of those parts were classified as "Criticality
1," meaning their failure would lead to what the agency dryly
described as "loss of mission, vehicle and crew."

Dick Scobee saw the risks in engineering terms. For him, spaceflight called for a calculation comparable to that of piloting a twin-
engine C-7 Caribou through enemy fire over North Vietnam: the
risk was real but acceptable. To his college-professor wife, history
mattered as much as math. For June Scobee, the space program
was a complex interplay of engineering and history. She knew
that in NASA terms, engineering *is* history. It was only after the
second Mercury mission in 1961, when Gus Grissom's capsule
splashed down in the Atlantic and the explosive bolts on its hatch

fired too soon, nearly drowning him, that the agency modified the bolts' design. The new design kept the hatch from opening accidentally. But it made the process take longer. In 1967, when a spark in the Apollo 1 capsule set off the fire that killed Grissom, Ed White, and Roger Chaffee, it took NASA workers five minutes to open the hatch.

Almost twenty years later, that was still the only fatal accident on the launchpad. No astronaut had ever died in space, though there had been close calls. Apollo 11's lunar module had come within fifteen seconds of running out of fuel as Neil Armstrong sought a landing spot on the moon. A year later, an oxygen-tank explosion ("Houston, we've had a problem") threatened to leave Apollo 13's Jim Lovell, Jack Swigert, and Fred Haise marooned in space. Improvisation and good fortune brought them home alive. Then, in 1984, Commander Hank Hartsfield, mission specialists Resnik and Mullane, and three others were strapped into *Discovery* when the launch countdown stopped at 00:00:04. The engines shut down. Alarms sounded; there was a fire on the pad. "Hank ordered us to unstrap and prepare for emergency egress," Mullane remembered. "Judy crawled to the side hatch window and reported the access arm had swung back into place and the fire suppression system was spraying water. She didn't see any fire."

Resnik pressed the intercom button on her flight suit and asked the commander, "Do you want me to open the hatch?" They knew that hydrogen fumes from the external tank and the shuttle's engines posed a particular threat. Unlike flames fed by propane or a carbon-heavy campfire, a fire fed by pure hydrogen is practically invisible—there are no combustion products to turn the flames red. If they left the shuttle for the slidewires that could zip them to safety—the escape they had practiced in training—they might slide right into flames they couldn't see.

Hartsfield said, "Negative on opening the hatch, Judy."

Mullane remembered it as "a decision that might have saved our lives." As it turned out, hydrogen leaking from one of *Discovery*'s engines had caught fire, scorching the shuttle from its tail all the way to the crew cabin. Opening the hatch could have killed one or more of them.

A year and a half later, with the Teacher in Space mission on the launchpad, the shuttle program had a perfect record: twenty-four missions and twenty-four safe landings, with no bruises much worse than the bump on Mattingly's head in 1982. Still, the space truck had its doubters. Gregg Easterbrook of the *Washington Monthly* called the shuttle "a deathtrap. During blastoff, there is no way out. Here's the plan: Suppose one of the solid-fueled rocket boosters fails. The plan is, you die." Engineers at Morton Thiokol, the contractor that built the rocket boosters, had noticed burn marks on some of the seals that held the boosters together. They shared their findings with colleagues, who promised to study them. Nobody shared those concerns with the astronauts.

Mullane claimed it wouldn't have mattered. "If Dr. Kraft [JSC director Christopher Kraft] had explained exactly what we had signed up to do—to be some of the first humans to ride uncontrollable solid-fueled rocket boosters without the protection of an in-flight escape system, on a launch schedule that would stretch manpower and resources to their limits—it wouldn't have diminished our enthusiasm one iota. For many of us, our life's quest had been to hear our names read into history as astronauts."

Christa's quest was different. "I can't wait to get home and get back to work," she told an interviewer. "If the Teacher in Space doesn't return to the classroom, something is wrong." Her contract with NASA called for her to spend a year traveling the country, promoting the space program, after *Challenger*'s weeklong mission. Then Mrs. McAuliffe would be free to return to room 305 at Concord High. Her husband had hinted to the press that she might run for mayor or even for Congress, but when Christa looked ahead to 1987, she saw herself returning to Steve and their children, teaching, lobbying for better pay for teachers, maybe applying for assistant principal. "I had no doubts that after her year of being the Teacher in Space she would teach a few more years," her mother wrote, "and then go on to a position in school administration. There she would be better able to help cure some of the nation's educational ills." But during her last days at the Cape, Christa wondered if she was being selfish to pursue her dream of making history while her six-year-old daughter cried herself to sleep at night.

"Would you do it?" she asked June Scobee. "Would you go up in the shuttle?"

"In a minute!" June said. She asked Christa to think of how proud her husband and kids were that she, Christa McAuliffe from little old Concord, New Hampshire, was the Teacher in Space. June asked her to picture all the boys and girls her space lessons would inspire. How many would go on to careers in science and technology? How many would want to be schoolteachers? In June's view, the Teacher in Space program exemplified one of Christa's favorite sayings: *I touch the future—I teach.* As for the risks, riding rockets was "scary at first," June said, "but it's like riding a bicycle. You get used to it."

"These were things I'd heard Dick say," June Scobee recalled years later. "I was just repeating them to Christa. We all believed them. So when she asked, 'Would you fly?' I said yes. I said, 'If it's safe enough for my husband, it's safe enough.'"

—◆—

The week before the launch, Scott McAuliffe and eighteen of his third-grade classmates flew to Florida along with thirteen of their parents. United Airlines gave them all free tickets. Sponsors including Adidas, Nike, and Puma supplied free clothing, shoes, and backpacks identifying the children as CHRISTA'S KIDS. The Amelia Earhart Luggage Company provided suitcases for them. Just before their plane took off, when a *Monitor* photographer showed Scott a photo of Christa in her flight suit and asked how he felt about watching the launch, he began to cry.

Scott perked up when the pilot announced that "Christa McAuliffe's son" was onboard. Passengers applauded. Deplaning in Orlando, lugging his Earhart suitcase and a plush space shuttle that was almost half his size, the boy made his way through a throng of reporters, TV crews, and well-wishers, plus Ronald McDonald in a spacesuit. One of his friends said, "Scott's getting tired of people following him around."

Per NASA tradition, shuttle crews' families and friends held a party on the night before a launch. Crew members couldn't attend—they were quarantined in their dorm rooms. With the launch set for Super Bowl Sunday, more than a hundred family members, in-laws and hangers-on gathered in a Holiday Inn ballroom in Orlando that Saturday evening. The Corrigans' table featured an easel displaying blowups of Christa as a baby, a Girl Scout, a bride, a teacher, and as NASA's most famous payload specialist. "A grand party," Grace called it. "Ed and I stood at Christa's table, Ed proudly wearing his huge Teacher in Space button with Christa's picture." Steve McAuliffe posed for photos and told everyone in sight that the Patriots were going to beat the Bears in the Super Bowl. He also signed a few autographs. When Caroline asked why anybody would want his autograph, Steve said, "Because I married Mommy."

10

"I DON'T THINK ANY TEACHER HAS EVER BEEN MORE READY," Christa told reporters. In the six months since Vice President Bush named her America's Teacher in Space, she had mastered more science than she ever expected to. She had learned how to bob weightlessly from one end of the Vomit Comet to the other. She had kept her wits together on national TV, promoting the space program as well as the cause of America's schoolteachers, fighting NASA when she felt she had to, while coaching Steve by phone to keep him from burning the house down. The February 1986 issue of *New Woman*, which hit newsstands that week, called her "America's most ordinary celebrity," a wife and mom "with the gift of gab to get people excited about the space program again."

She was packed and ready to fly. Before leaving Houston for

Cape Canaveral she put together her one-and-a-half-pound personal preference kit (PPK), a nylon pouch filled with items that would travel to space with her. Her PPK held Steve's VMI class ring; Caroline's gold crucifix necklace; Scott's toy frog, Fleegle, vacuum-packed with its stuffing removed to save space; a Girl Scout pin; miniature flags representing Concord High School, Framingham State College, Marian High School in Framingham, and the National Educational Association; a copy of the poem "High Flight"; and a pair of T-shirts, one with the state seal of New Hampshire and one reading *I Touch the Future—I Teach.* She was also allowed to take a cassette player, her favorite Bob Dylan and Carly Simon tapes, and a laserdisc provided by NASA's corporate sponsor 3M, holding "signed pledges from hundreds of thousands of children throughout the world to work for peace now and as a legacy for future generations . . . reduced to space-travel compactness."

The agency's meteorologists in Houston and Florida were calling for a fifty-fifty chance of rain on Super Bowl Sunday. That prediction troubled the forecasters, who pointed out that raindrops can "act like bullets" at the speeds a shuttle reaches on the way up, possibly damaging the silicon tiles that would protect it during reentry. Mike Smith wanted to fly anyway. Smith could cite chapter and verse about rainy-day missions that went off without a hitch. Apollo 12 had taken off in a storm and climbed right through a bolt of lightning. "I'm willing to go with a fifty-fifty chance of rain, and so is Dick," Smith told the forecasters. "I'm a hell of a lot more concerned about an approaching cold front than some rain that may or may not materialize." But the commander and pilot would not take charge of the shuttle until it was two hundred miles off the ground. Fearing rain, the agency pushed the mission back another day, to Monday the twenty-seventh.

The latest scrub disappointed thousands who had traveled to Florida to see Christa blast off. They included Vice President Bush as well as former payload specialists Garn and Nelson, who flew back to Washington. Dozens of the crew members' relatives and friends headed home to avoid missing work or school, but thousands more remained at the Cape—more spectators than for any launch since Sally Ride's.

Slipping *Challenger*'s launch to Monday made sense from a weather standpoint. More important, perhaps, it delighted NASA's public affairs officers. Launching the Teacher in Space on a weekend would have been a disaster for them. Only a weekday launch could be televised live to two and a half million schoolchildren. As astronaut Mullane put it, "To the surprise of no astronauts, NASA went to work to revise the flight plan and move the space lesson to a school day."

—◆—

Smith woke early on Sunday. He opened his blinds in the crew's quarantine quarters and got a face full of sunshine: perfect flying weather. They could have been on their way uphill long before the cold front stretching south from the Arctic reached the Cape. Leave it to landlubbers to let politics and PR affect engineering decisions. "You've got people making decisions down here who've never even flown an airplane," he griped. As John Young, the first shuttle commander, liked to say, "Ain't none of them boys ever died because a desk crashed."

The rain held off until Sunday afternoon. NASA spokesman George Diller had good news for reporters gathered in the domed press center at KSC: Monday's forecast called for clear skies, though the weather would be cooler. That was the good news. The bad news was that Tuesday looked worse. "The weather will turn much colder," Miller said, "with a low temperature forecast of twenty-one degrees in Orlando. This would be a constraint to launch."

Smith warned his in-laws that the mission might be scrubbed again on Monday. If that happened, given the record-setting cold that was on its way, "there's no chance we'll fly till Thursday."

With the crew members quarantined, many of their friends and close relatives joined Christa's parents for a Super Bowl party. "It's beer for the Patriots today and champagne for Christa tomorrow," Grace Corrigan declared. For the Patriots, Super Bowl XX went south faster than the weather. In the third quarter, the Bears' 335-pound lineman William "Refrigerator" Perry rumbled to a touchdown to put Chicago ahead 44–3.

Christa and her crewmates watched New England's loss from their quarters in the O&C Building. Spouses who had passed a medical test to prove they weren't infectious were allowed to join them. Marcia Jarvis got a surprise from husband Greg, who had turned the delays to good use by working on his long-delayed MBA thesis. Handing her a manila envelope, he said, "I finished my paper!" After coffee and dessert the McNairs brought out a three-liter jeroboam of Moët & Chandon champagne and an engraving pen. All seven crew members etched their signatures on the bottle. "After we land, we're going to have one heck of a party," Ron McNair announced, "and this champagne will kick it off."

Lights-out was at 7:00 p.m., giving the Challenger Seven more than nine hours until their Monday-morning wakeup calls. They ignored the curfew. The Super Bowl went on past their supposed bedtime, and nobody wanted to sleep. By then Christa had already gone down the hall to tap on Greg Jarvis's door. Did he feel like going for a walk? He sure did.

They had barely started when they came across a pair of delivery bicycles. *Challenger*'s payload specialists commandeered the bikes and pedaled toward the launchpad. A TV crew happened to be going the other way on State Road 3. The producer recognized the Teacher in Space and had his crew set off after her, filming her on the eve of her big day.

Christa smiled and gave them a wave. "Must be a slow news day," she said. "Don't get too close—we're in quarantine!"

———◆———

Technicians began fueling the stack at 3:18 a.m. on Monday the twenty-seventh. It would take three hours to pump 535,000 gallons of supercooled, supercombustible liquid hydrogen and oxygen through aluminum fuel lines into the giant external tank, giving the tank the explosive power of a nuclear bomb. Liquid oxygen has a temperature of minus 297 degrees Fahrenheit and a pale blue tinge. Liquid hydrogen is colorless and colder still, colder than Pluto at minus 423, only 37 degrees above absolute zero. Neither substance is found in nature. Both are far colder than ice, yet once mixed and ignited in the orbiter's engines they produced the 6,000-degree fire that would help drive the shuttle skyward. Oddly enough, the by-product of their combustion was water, and as countless hydrogen atoms combined with oxygen to form superheated molecules of H_2O, the shuttle's exhaust was great white billows of steam.

Christa woke a little after five o'clock on Monday morning. She showered, tugged on a pair of jeans and a white polo shirt, and joined the others for the traditional prelaunch breakfast. Their matching NASA polo shirts were emblazoned with the official patch of shuttle mission 51L, showing *Challenger* zipping around the earth with the American flag waving in the background while Halley's Comet approached. The crew's seven surnames circled the image, with a small red apple beside MCAULIFFE, the Space Teacher.

The Challenger Seven took their places at a breakfast table that held a pot of red roses, white carnations, and miniature American flags. They tried to make small talk while photographers snapped pictures. The prelaunch breakfast menu never varied: steak and eggs, a custom dating back to Project Mercury. They picked at their food—all except Resnik, who dug in. After the meal, servers brought out a large white cake decorated with the mission patch rendered in frosting. Scobee and his crew sent compliments to the kitchen, but not even Resnik took a bite. By KSC tradition, tasting the launch-morning cake was thought to be bad luck. Vowing to enjoy it when they got back, they left the cake untouched.

After breakfast they had a half hour to themselves. They made a few brief, final phone calls, then went downstairs to change into their blue flight suits and go out to meet the Astrovan, which would ferry them to the launchpad. The ride to the launchpad would take twenty minutes, but they couldn't leave just yet. Another launch-day custom decreed that the van could not depart until the commander won a hand of cards. The game was a form of lowball poker, an old test pilots' pastime called Possum's Fargo: five cards, no draw, no betting, and the worst hand wins. Scobee stood around a high table in the suit-up room in the O&C Building, playing the game with George Abbey and another NASA official. Only after the commander won a hand could he lead his crew out the door.

They emerged at six forty-five, waving to roped-off rows of photographers and TV crews on their way to the Astrovan, a silver 1983 Airstream mobile home that would have looked right at home on a KOA Kampground. Wood-paneled on the inside, with blinds on the windows, the Astrovan carried them on its nine-mile route along an access road where alligators often sunned themselves in better weather. The road ended at Launchpad 39B, where *Challenger* stood waiting for its crew.

No astronaut ever forgot the moment he or she first reached

the pad on launch day, with the stack—the orbiter mated to its mammoth fuel tank, flanked by twin rocket boosters—held in place by the 380-foot tower. "That's the first time you ever saw a living breathing orbiter," recalled Hoot Gibson, who considered the shuttle a living being "because it had been fueled. You had the liquid hydrogen and liquid oxygen in the tank. The thing was hissing and clanking," with vents releasing excess fuel that turned instantly to vapor. "It sounds like it's breathing and it's alive . . . hissing and breathing and clanking and creaking."

They rode the tower elevator fifteen stories to the level of the White Room, a sterile compartment with the ambience of a jetway, where KSC's closeout crew waited to prep the crew for flight. To reach the White Room from the tower elevator, they had to walk across a steel catwalk.

Christa's stomach lurched at the prospect of crossing the catwalk 150 feet above the ground. Resnik told her not to worry—she'd felt a little dizzy her first time, too. "You'll be fine."

The catwalk's floor was painted yellow, with directional chevrons pointing away from the shuttle. The closeout crew called it the "Yellow Brick Road." The chevrons were there in case a prelaunch emergency sent them running the other way. Such an emergency was likely to be a fire that set off the launch tower's powerful sprinklers, drenching them all in such torrents of water that they could only see their feet. Like the slow-opening hatch, the sprinklers and Yellow Brick Road were safety features that predated the shuttle program.

They had rehearsed this, but not with the stack huffing and puffing. "I'm not sure how I'll feel," Christa had said. "It's kind of like the first time you got on a carnival ride. You said, 'I've got enough courage,' and you're excited about conquering your fears." She made it across.

The White Room, named for its spotless interior, was the size of a walk-in closet. It could accommodate only two or three crew members at a time. Scobee and Smith went first, joshing with men in white jumpsuits who helped the commander and pilot don their egress harnesses, which would hook to a cable at the end of the Yellow Brick Road if they had to make a run for it.

Finally, the closeout crew wiped down the soles of their boots, and Scobee and Smith crawled through the circular hatch into the orbiter. With *Challenger* aimed straight up in launch position, everything inside was sideways, with seats bolted to a vertical floor. They crept through the dimly lit middeck to a ladder that led to the flight deck, where sun streamed through the cockpit windows. Scobee strapped into the left-hand seat while Smith strapped into the pilot's seat to his right, both with their backs to the earth and knees above their heads. They took the day's first readings from some of the 1,300 switches, gauges, and TV monitors arrayed on control panels in front of them, between them, and on the ceiling above them. Members of the closeout crew, working with the efficiency of a NASCAR pit crew, attached air hoses to their helmets, connected their comm lines, checked their microphones and earphones.

Resnik and Onizuka were next to enter the White Room. They clambered through the hatch and climbed "up" the horizontal ladder to the flight deck, where they strapped into their seats behind Scobee and Smith. As the flight engineer, Resnik would be busy during the ascent. As one astronaut wrote, "The position of flight engineer was arguably one of the most important jobs a mission specialist could possibly hold. Seated behind and between the commander and pilot during the critical periods of ascent and reentry, he or she was responsible for helping monitor the orbiter's instruments and offering a vital third set of eyeballs in the event of 'off-nominal' events."

"Off-nominal" is engineer talk for surprising.

In the White Room, Christa chatted with the closeout crew while she and Jarvis put on their harnesses and helmets. They lifted their shoes for the soles to be cleaned. At 7:36, she clambered through the hatch to her armless steel seat on the second deck. Like the other seats, hers featured thin, dark blue cushions, but was less comfortable than an airline seat.

With more than fifty pounds of equipment on her, including her flight suit, long underwear, boots, gloves, helmet, and egress harness, Christa needed a hand strapping in, as all astronauts did. Sonny Carter, a rookie astronaut serving as ASP (astronaut support person), helped her with the seat belts that came over both shoulders to meet the belt around her waist. Carter helped connect her hoses and comm lines.

McNair settled into his middeck seat a few feet away, beside the open hatch. Carter and another worker helped him strap in, then said their goodbyes. They shook hands with McNair, crawled out, and began closing the hatch door.

Now came a moment no simulator could mimic. During months of sims and onsite rehearsals, the hatch never closed. Once it clanked shut, the crew was alone in the orbiter for the first time. This was no drill; it was "go time."

McNair had the best seat on the middeck. "His spot was like the flight attendant's jump seat on a plane," says former training chief Frank Hughes. In case of a launchpad emergency, McNair would open the hatch and help the others hook their egress harnesses to the cable that would zip them to safety. For now, he, Christa, and Jarvis lay on their backs, looking up at the floor of the flight deck, where the other four enjoyed a panoramic view of the morning. McNair at least had a porthole in the hatch door to look through. The ten-inch window allowed him to watch Johnny Corlew, a former navy mechanic who had served three tours repairing fighter jets during the Vietnam War, tick boxes on his operational maintenance instruction (OMI) checklist. Inspection technician Corlew was moving down the list when he reached a line having to do with a pair of switches inside the orbiter. The switches should indicate that the hatch is *latched and locked.* One did, but one didn't. Scobee reported that he had an error message from the hatch. Another message said the hatch was locked; they were good to go. But the countdown could not continue with an error message flashing.

Was the hatch locked? The porthole in the hatch door allowed Corlew to see the locking mechanism. He used a lighted mirror to get a better look. The door looked to be latched and locked as it should be. The error message was probably a glitch, but it was his job to double- and triple-check. Better to burn a few minutes or even scrub a launch than to be wrong about a Criticality 1 component. A leaky hatch could cause a catastrophic decompression in the orbiter. The crew might die on the way up.

Corlew radioed launch director Gene Thomas in the Firing Room, a tiered amphitheater in the Launch Control Center where Thomas presided over more than a hundred engineers sitting at computer consoles. Winter winds off the Atlantic were picking up speed. The longer they waited, the greater the chance of another embarrassing scrub. Thomas dispatched a team of technicians to the launchpad. With McNair watching through his porthole in the hatch, the new crew tried using a T-handle wrench on the door. They spent half an hour trying to loosen a bolt in the locking mechanism. They succeeded only in stripping the bolt. Corlew radioed: "Send us a drill."

Forty-five minutes later, a maintenance van pulled up at the pad with a cordless Black & Decker drill. Unfortunately for Corlew and his team, the drill was out of juice. They sent for replacement batteries. That took another half hour. Christa, strapped into her seat on the middeck with Jarvis and McNair, did her best to relax. Their launch window was closing as the cold front Smith had been tracking for a week moved toward Cape Canaveral. As the delay ran on, the forecasters said they had an hour, maybe two, before the winds got too strong to launch.

The replacement batteries were dead. "That's when I started having misgivings," recalls *Monitor* reporter Ralph Jimenez. "It was almost like a premonition. They can't get batteries for a cheapo cordless drill?" Other reporters compared the hubbub to a *Three Stooges* episode. Finally Corlew radioed Launch Control, asking for permission to try another device: a hacksaw.

The hacksaw worked. They sawed the bolt loose and discovered that Scobee's first suspicions had been correct: the hatch was latched and locked the whole time. The problem was an error-message malfunction.

NASA workers had a name for such snafus—"instrument funnies"—but *Challenger*'s usually upbeat crew wasn't laughing. They crawled back out through the hacksaw-scraped hatch, stiff and sore, feet and hands numb, heads groggy from the blood pooling in their brains as they waited out a five-hour delay. With thirty-mile-per-hour winds whipping flags at the Cape, the launch was scrubbed for the sixth time.

Scobee said he'd had a bad feeling about their chances when he heard the sound of drilling coming from the middeck. "It sounded like a bunch of termites." Corlew apologized for ruining his day. Scobee said, "Forget it, Johnny. We'll try again tomorrow."

The traditional lunch of beans and cornbread—KSC workers' reward for a successful launch—would have to wait another day. Launch director Thomas was heading home for a few hours' sleep when someone handed him a Xerox copy of a ditty that was going around the Firing Room. A parody of "The Super Bowl Shuffle," the Chicago Bears' jokey music video, it was one of what Thomas called "the many comical writings" by KSC engineers:

We missed the good weather when the hatch wouldn't close
Then we messed up with the dead drill we chose...
First we held the Congressman, and then we held the Teacher
But holding to schedule is not our main feature.
We all hate to scrub because we know what that means
The Super Scrub Shuffle and a wave-off on beans!

—◆—

All three network news programs featured NASA's latest embarrassment. CBS anchor Dan Rather called "today's high-tech low comedy" a fiasco, "yet another costly, red-faces-all-around space shuttle delay." As ABC's Peter Jennings saw it, "Once again a flawless liftoff proved to be too much of a challenge for the *Challenger*."

Mike Smith couldn't believe they weren't uphill already. "We'll get there," he told his wife. Smith phoned his mother-in-law to tell her she might as well go back to work for a couple of days. Given the size of the cold front on his weather charts, "There's no way we'll fly until Thursday."

Resnik was more upbeat. Later that day she spent a few minutes with Mullane, her *Discovery* crewmate. When he asked how tomorrow looked, she said, "Good, except it's supposed to be cold, down in the twenties. We're worried about ice." Still, she was counting on a Tuesday launch that would return them to Earth in a week. "I'll see you back in Houston," she said.

The families gathered for yet another prelaunch dinner. Steve McAuliffe admitted being frazzled but told reporters he didn't mind waiting out so many delays. "I have found them very comforting," he said. "It seems to me that it demonstrates beyond anyone's normal patience that they are not sending them up unless everything is perfect."

That night, Barbara Morgan tapped on Christa's door. She

found the Teacher in Space hard at work in her quarantine dorm room. "She was such a real, hardworking teacher that in the crew quarters at Kennedy Space Center, on the day before lift-off, she was writing college recommendations for her high school students."

On the phone to her sister Lisa, Christa admitted she was feeling "some butterflies" about the launch delays, but said she had "full faith in NASA."

The weather got worse. A National Weather Service bulletin warned Floridians of "a major freeze. . . . Temperatures may approach all-time record lows." Citrus growers set bonfires to protect their oranges and grapefruit from the subfreezing temperatures to come. NASA technicians kept the sprinklers up and down the launch tower trickling overnight to keep the water lines from freezing. They added antifreeze to the water that fed the water lines.

"Ron got the word from NASA: You can sleep in tomorrow. It's going to be too cold," one of McNair's friends recalls. "Then he heard a different story: 'The launch is back on.'"

Christa phoned her mother. "No matter what, Mom, we're going tomorrow," she said.

11

O VERNIGHT TEMPERATURES FELL TO 22 DEGREES. LAUNCH-
tower sprinklers gave the tower a coating of water. Technicians
added antifreeze to the water in the trough under the stack. The
water froze anyway. Soon the tower's railing, cameras, and sprinklers were dripping with icicles.

A fueling glitch led to the day's first delay. Christa and the
others got an extra hour to "sleep in" until their bedside tele-
phones rang at 6:18 on Tuesday morning. For the second day in
a row they had half an hour to shower, dress for breakfast, and
make a phone call or two. Scobee phoned his wife. "Good morning, sleepy," he told June. "It's a great day to fly."

"In these freezing temperatures?"

He said she had nothing to worry about. "They're knocking
off the icicles. Well, I have to go now. Tell the kids I love 'em, and

I'll see you in a week. I'll miss you. Bye, honey!" He closed with a kissing sound, the way the Scobees ended phone calls. June kept the receiver to her ear until there was only a dial tone.

—◆—

It was still dark outside when flight director Randy Stone's phone rang. "I'm waking you up because it's freezing at the Cape," night-shift director Chuck Knarr told Stone. "I don't think we're going to go. There's ice on the pad."

Stone went through a mental checklist. "Is the Control Center ready?"

"Yes."

"Is the vehicle—the parts we can see and are responsible for—is it ready to go?"

"Yes."

"Well then, from an operations standpoint, we are go for launch."

—◆—

For the second day in a row, the crew convened for a prelaunch breakfast. Again they picked at their steak and eggs, with Resnik reaching for seconds. Again they oohed and aahed over a cake frosted with white icing and the mission emblem—a freshly baked duplicate of the first cake—and promised to eat it later.

After breakfast they filed into a conference room for a weather briefing. The weather was cold but improving, their briefer said. The temperature was up to 24. That was far colder than for any previous shuttle launch, but was not by itself a launch constraint.

Once again they suited up, waited for Scobee to win a hand of Possum's Fargo, and paraded out of the O&C Building for the traditional astronaut walkout, a brief photo op in which they waved to reporters on their short march to the Astrovan. Scobee led the way, followed by a smiling, waving Resnik. Next came McNair, looking forward to his last space mission, and Smith, rubbing his hands together to keep warm while looking forward to his first mission, followed by payload specialist McAuliffe, falling in step behind Smith. "Go today!" she said. Onizuka and Jarvis followed her into the Astrovan, where Onizuka was glad to find the warm jacket he'd arranged to have waiting for him. There was a dress code for the walkout: flight suits only. The happy Hawaiian bundled up as soon as they boarded the van.

—◗◆◖—

Steve McAuliffe rousted Scott and Caroline from their beds in a Disney World hotel and led them to the parking lot. He would remember scraping frost off the windshield of his rental car that morning. At the Cape, where his VIP badge got them into the Launch Control Center, they took the elevator to the fourth floor and joined the other crew members' families in launch director Gene Thomas's offices. The windows gave them a wide view of KSC with the launchpad in the distance. Thomas's secretary brought out a sheet of drawing paper and Magic Markers for the kids. The grown-ups chatted over coffee and doughnuts while the children drew or dozed or played tag, a couple of the older ones reading magazines, one of the youngest sucking his thumb. "Then there was Scott, Christa's boy. He didn't want to leave the window," June Scobee recalls. "Scott kept his nose to the glass, looking out at the shuttle."

Outside, Christa's parents and other relatives and friends filed into the VIP bleachers nearby, many using binoculars to peer across three and a half miles of swampland to the stack on its launchpad. Before them stood the space center's stadium-sized countdown clock, with its digital display showing the hours, minutes, and seconds until launch in numbers four feet high and two feet wide, counting down from -04:00:00 to -03:59:59. The

Monitor's Ralph Jimenez remembers "a carnival atmosphere—except for the cold." CNN's John Zarrella looked up at a sky the pale blue of liquid oxygen. "There wasn't a cloud. You could cut that sky with a knife." The 113 Teacher in Space semifinalists were there, some bundled in blankets, some jogging in place to keep warm. All of them had received sky-blue flight suits like the crew's and planned to wear them on educational tours in their home states, promoting NASA and the space-participant program after Christa and *Challenger* returned to Earth. They applauded when Scott McAuliffe's classmates from Kimball School in Concord unfurled a banner: GO CHRISTA.

Christa's parents, her sister Lisa, and her brother Chris, known as Kit, sipped from steaming cups of coffee and hot chocolate. Ed Corrigan wore a warm cloth hat and a scarf over a topcoat that held a pair of lapel buttons showing his daughter's smiling face. Grace, dressed in a fur-collared coat and leather gloves, recalled their hours in the bleachers as "cold, cold, cold." Despite the chill, "The word was out that today was the day."

—◆—

The Astrovan reached the launchpad at 8:03. Scobee brought a gift for the closeout crew, a two-dollar bolt like the one that had caused the previous day's scrub. "You guys might need this today," he said.

Crew chief Johnny Corlew said, "It's a pretty cold day to be flying."

"No, it's great. Nice and clear," Scobee said.

Maybe so, Corlew told him, but if Commander Scobee and his crew had any interest in using the toilet before boarding, they should go back to the one in the Astrovan. The toilet at the launchpad was frozen.

Again the commander and pilot climbed into the orbiter.

They crept sideways across the ladder to the flight deck, where the support team's Sonny Carter helped strap them into their seats. Scobee tapped his comm link and heard from launch director Thomas. "Welcome to our northern launch facility," Thomas said. Launchpad 39B was a mile and a half north of 39A, site of all twenty-four shuttle liftoffs to that point.

"Ha. Thanks a lot," Scobee said.

"The northernmost pad we've got," Thomas called it. "You can probably tell from the icicles how far north it is."

Christa was crossing the yellow catwalk to the White Room. She and Resnik and McNair and Onizuka entered together, shivering, stamping their feet against the cold. Corlew had a gift for her: a red apple for the teacher.

"Save it for me," Christa said. "I'll eat it when I get back."

Corlew's crew helped Resnik secure her flight helmet. Black curls spilled out from under the back of the helmet. Resnik gave the closeout crew a thumbs-up. She followed Onizuka into the shuttle—but first she turned to Christa.

"The next time I see you," she said, "we'll be in space!"

It was warmer in the orbiter than in the White Room but still cool: 61 degrees. Resnik moved smoothly to the flight deck and strapped into her jump seat behind Scobee and Smith. Onizuka, sitting beside her, said, "My nose is freezing."

Resnik's vantage gave her a view of clear skies in the forward windows. From the flight engineer's seat she could reach panels of overhead switches the others couldn't. For the second day in a row the closeout crew handed her an eighteen-inch "swizzle stick," an aluminum rod that would allow her to reach those switches even if G-forces pinned her to her seat. At three Gs, the maximum expected on the uphill to orbit, her arm would feel three times heavier than normal. It would take strength and focus to use the stick to throw switches at three Gs, not to mention the higher forces they might face in an emergency.

Lying on her back in the flight deck's second row, facing the shuttle's front windows and the sky beyond, Resnik was where she wanted to be. In all the world there was no better seat, except for the commander's and pilot's positions, and who knew where the second female astronaut in NASA history might be stationed on a future flight? No mission specialist ever graduated to shuttle pilot, much less to the commander's seat. Flyers and scientists occupied different career tracks, with flyers like Scobee and Smith in the orbiters' front seats while specialists like her played support roles. But if anyone could imagine breaking the barrier between the flight deck's front and back rows it might be Judy Resnik, the best jet pilot who ever published a paper in the *Journal of the Optical Society of America*. With a little over an hour to launch—she hoped—Corlew and his team checked her air and comm lines.

"Good morning, Judy." A voice in her headset, checking her comm link. She was supposed to answer "Roger" or "Loud and clear." But after seven and a half years in the space program, after a thousand slights, including an astronaut's salary of $50,004 that was a fraction of what she could earn in the private sector, after a thousand reporters' questions about makeup, hairstyles, and sex in space, Resnik was fired up.

"Cowabunga!" she said.

Scobee liked that. He said, "Loud and clear there, Judy."

With the flight deck full, all systems go, Sonny Carter went to the middeck to help Christa plug in.

"Christa, you ought to be able to hear me," he said, checking her radio link.

"Real fine," she said.

"Okay," Carter said. "Talk to the OTC." Orbiter test conductor Roberta Wyrick was waiting for her signal at Launch Control.

Christa pressed the "transmit" switch connected to her headset. "OTC, PS-1," said payload specialist McAuliffe.

"Loud and clear," said Wyrick, adding that she hoped this would be their go day.

"Good morning," Christa said. "I hope so, too."

Once her comms were in order and her straps and helmet snug, Carter helped Greg Jarvis strap into his seat. Jarvis was PS-2, a bit of a postscript on the Teacher in Space flight, not that you could tell it from his comm check. Asked how he was feeling, he yelled, "Faan-tastic!"

The middeck flyers laughed. After half a dozen scrubs, after being bumped from two other flights by a senator and a congressman, good old Greg was raring to go.

They were all in their seats by 8:40. Liftoff had been scheduled for 9:38 pending a last inspection by the ice team, a crew of specialists in orange coveralls who checked the tower for icicles that might crack off during the launch and damage the heat-resistant silicon scales that protected the orbiter's hide. The ice team finished its inspection at 8:44. Its verdict: there was too much ice on the pad. Ice team chief Charlie Stevenson radioed his bosses: "The only choice you've got today is *not* to go."

—◆—

In Downey, California, Rockwell International executives and engineers reported to work before dawn to watch the liftoff on NASA's closed-circuit TV channel. Rockwell built the orbiters and was part of the agency's command-and-control chain. Rocco Petrone, the company's president of space-shuttle operations, was a headstrong former NASA man, a former director of the Apollo program. With his dark, thinning hair and caterpillar eyebrows he resembled the *Saturday Night Live* comic John Belushi, with a growly voice to match.

Petrone didn't like what he saw on NASA TV. During an early-morning meeting with Rockwell engineers, he heard one say the launchpad "looks like something out of *Doctor Zhivago*. Sheets of icicles hanging everywhere. The big concern is that nobody knows what the hell is going to happen when that thing lights off and all that ice gets shook loose and comes tumbling down—what does it do then? Ricochet? Does it get into turbulent conditions that throw it against the vehicle?"

Petrone phoned Bob Glaysher, Rockwell's representative at KSC. He said, "Make sure NASA understands that Rockwell feels it is not safe to launch."

After hearing from Glaysher, Arnold Aldrich, NASA's shuttle-program manager, agreed to a delay. "As we marched to launch," Aldrich recalled, he and the agency's other decision makers decided to "let the day get warmer. We were originally going to launch around nine. We slipped it to almost noontime, and things were warm and melting."

Ice team workers swept icicles off the stack and tower. Some of the icicles were two feet long. The cold had knocked out a hazardous-gas detector and a video camera on the launchpad. A technician pointed an infrared camera at one of the shuttle's rocket boosters, which had spent the night in subfreezing winds and fumes from the supercooled external tank. He got a reading of nine degrees.

Scobee got an update from Launch Control. Another delay.

"Commander understands a T-equals-zero no earlier than eleven oh-eight," he told his crew on the intercom. T stood for time: "T minus ten" was the beginning of the "Ten, nine, eight . . ." countdown every schoolchild knows. "T-equals-zero" meant launch. Scobee was telling them they were in for a wait of at least two hours. "Everybody hear that?"

"Unfortunately," Jarvis said.

Onizuka joked about the cold: "Gunga Din doesn't know how to operate in cold weather."

Smith had spent a week being a good soldier, watching the weather by the hour and the minute. Now it appeared he was right to say they might not launch until February. "I feel like I'm at the four-hour point of yesterday," he said.

"I feel like I'm past it," Resnik said. "My butt is dead already."

Mission specialist Onizuka offered to massage her butt back to life. That got laughs over their comm lines. Smith nominated Onizuka as "crew gynecologist."

For Christa this was a little like eavesdropping on the cool kids at school. Jarvis might chip in with a joke, but in general the middeck crew kept quiet while the flight-deck crew did the talking. She heard Onizuka whoop with pleasure when the close-out crew sent a blast of warm air through the flight deck. Resnik reported that the hot air "went right up my you-know-what."

They waited. An hour passed as the four on the flight deck watched what appeared to be snow fall past the forward and over-head windows. In fact, they were seeing frost drift off the super-cooled external tank.

Resnik said, "I hope we don't drive this down to the bitter end again today."

"We should have slept another hour," Scobee said.

Jarvis clicked in. "They're probably making a fortune selling coffee and doughnuts in the viewing area."

Sprinklers at the pad were still soaking the tower. Resnik said, "All those trickles they're trickling at a hundred gallons a minute are adding up."

"Where are they getting all that water?" Onizuka wondered.

"Your tax dollars," Resnik said.

Christa sat on her back on the middeck, elbow-to-elbow with

Jarvis, looking up past her knees at a bulkhead. The middeck view was famously drab. Most of their field of view consisted of a wall of white rectangular storage lockers. To Christa's right were the crew's spartan bunks. To her left was the galley, an aluminum cabinet they would open outward once they were in orbit to reveal a microwave oven, trays (magnetized to stick to a table or wall), food pouches, a water injector to hydrate freeze-dried items, drink containers (pouches with straws attached), and scissors to open the pouches. Velcro strips mounted on walls and other surfaces would help the crew keep Velcro-wrapped tools from floating around the cabin once they were in orbit. McNair sat behind Christa and Jarvis, peering through the ten-inch porthole in the hatch, the middeck's only window. He could see the frost drifting off the external tank, but said nothing.

Christa, too, kept her thoughts to herself. She wasn't a joker like Resnik, Jarvis, and Onizuka, and had no role in prelaunch preparations. While the flight deck beeped and buzzed with activity, the dim middeck was a waiting room. The middeck crew would have no role to play until *Challenger* reached orbit. At that point, however, payload specialist McAuliffe would be the focus of the mission. Perhaps she was thinking of what to say on her first transmission from space. She had often been asked about that. "I don't want to say anything phony or contrived," she'd told a reporter. "I want my perceptions to be honest, as natural as possible, and if they're ordinary, that's okay. Maybe that's just me." Maybe she would stick to the lines she used in a videotaped rehearsal: "Good morning. This is Christa McAuliffe, live from *Challenger* . . . on a field trip." Until then she had nothing to do but wait and listen to the chatter in her headset.

The others killed time with talk about their tingling limbs. "After sitting like this, I can't imagine how anybody can hang in those gravity boots," Resnik said.

"They don't do it for four hours!" Scobee said.

"Some people do. I saw one guy do about a hundred sit-ups from vertical—all the way up like you'd normally do a sit-up, but vertical."

Jarvis sounded impressed. "That takes some very, very strong abdominal muscles."

"Takes some pretty weak, weak brains," Resnik said.

——◆◆——

The shivering Corrigans watched through binoculars from the VIP bleachers. Seeing the ice team using broomsticks to knock icicles off the tower, "We had a very uneasy feeling," Grace recalled. "Things didn't seem to be going the way they should."

"I'd take her off that thing if I could get out there," Ed said.

But Grace knew their daughter. "Even if you could, she wouldn't come."

——◆◆——

The ice team completed its final sweep of the launchpad at eleven fifteen. Moments later, Scobee heard from Launch Control: "We're planning to come out of this hold on time." After six months of training, after six scrubs including yesterday's hatch-bolt follies, and after today's two-hour delay, *Challenger* was cleared to launch. Barring another last-minute scrub, they were minutes from liftoff. Scobee radioed back: "All riiight! That's great."

Their immediate families hurried to a wooden platform on the roof of the Launch Control Center, where they would have the best view of the launch. "Finally the long-awaited countdown was about to begin," June Scobee recalled. "We picked up the babies and cameras and climbed the stairs to the rooftop." Cheryl McNair took her three-year-old son's hand while speakers carried the countdown narration of Hugh Harris, the voice of Launch Control: "One minute away from picking up the count for the final nine minutes."

"My children and I stood with Steve McAuliffe and his children," June Scobee recalled. "The shuttle was beautiful against the clean blue sky."

—◆—

The pace of the flight-deck chatter picked up. "Mike, your boiler pre-act is next," Resnik said, referring to a preliminary stage of activating the shuttle's power units.

"Boiler pre-act complete," he said a moment later.

After more pre-checks, Resnik said, "My butt's going to like zero G a lot better than these seats."

"Mine, too," the commander said.

"You may or may not get a dp/dt," she said, reminding him of a cabin-pressure reading. "You might also get the high-press alarm as the cabin expands when we go up."

"Okay, Judy," Scobee said. "Seven minutes."

Two minutes later, Smith activated the shuttle's auxiliary power units, throwing three switches in the sequence he had practiced more times than he could count. "APUs coming on."

Scobee checked gauges on his side. "Pressure on all three APUs," he said. The orbiter was coming alive. Launch director Thomas radioed the flight deck. "We are going to give you a ride today," Thomas said. He checked their comm links by taking a roll call of the crew: "Can you read me?"

"Loud and clear," they said one after the other. Soon it was Christa's turn. "Loud and clear!" she said.

They sealed their helmets for launch. "Visors coming down," Scobee reported.

The countdown clock moved from -00:02:00 to -00:01:59.

"Welcome to space, guys," Scobee told his crew. "Everybody strap in tight. We're about to go for the ride of our lives."

—◆—

The morning had a pep-rally feeling at Concord High School, where balloons rose to the ceiling in the auditorium. Students and teachers gathered around TV sets there and in the cafeteria and dozens of classrooms. Some viewers wore party hats and blew on noisemakers. Students with kazoos played "Charge!" Others hoisted a banner: TO BOLDLY GO WHERE NO TEACHER HAS GONE BEFORE. Principal Foley and football coach LeBrun were enjoying a brighter day for Concord High than the Tuesday eight weeks before when Louie Cartier brought his shotgun to school. Eileen O'Hara, Christa's substitute, wore a Christa button as she sat by a TV set surrounded by teachers, students, and reporters.

In the White House, First Lady Nancy Reagan watched the countdown in the East Wing while her husband went over the latest draft of his State of the Union speech in the West Wing. The speech would give President Reagan a chance to highlight his vow to send "one of America's finest—a teacher" into space. By the time he spoke to both houses of Congress that evening, Reagan might be able to look skyward and salute Christa as the shuttle passed overhead.

At KSC the countdown clock read -00:01:30—ninety seconds to launch. Launch director Thomas radioed the flight deck: "Have a good mission." Scobee said, "Thanks a bunch. We'll see you when we get back."

As the sun rose higher over the Cape the temperature reached 36 degrees. According to one of NASA's internal documents, only three of the twenty-four shuttle missions had launched when the temperature was less than 60 degrees. The average of the two dozen previous launches was 72.8. That made this the coldest launch by at least 15 degrees, almost 40 degrees colder than the average.

The payload specialists leaned back in their seats on the mid-deck. Dim sunlight filtered in from the flight deck, where all the action was. Christa had no last-minute tasks. She had no way to converse with Jarvis, sitting directly beside her, except by intercom, and this was no time to clutter the intercom. She had nothing to do but pray there wouldn't be another delay. Barring some last-minute surprise, she was seconds from the violence of launch and less than ten minutes from the airless quiet of orbit—all within a year of the frazzled morning on the last day of the previous January when she hurried to the post office with her Teacher in Space application. Twenty-five years after watching Alan Shepard's suborbital flight from a middle-school cafeteria she was about to be part of history, her subject. She was about to join a line of pioneers tracing back to Shepard and the other original astronauts, the Mercury Seven, a line she hoped would continue to generations of "ordinary people" who would follow her into space. As she'd put it in one of her interviews, "The people who went to the moon, because they were able to see the whole world as this globe, came back with a much better perspective of why we all should work together. Because we are pretty fragile, when you look at the whole universe. So maybe as more and more people get that perspective, things are gonna change."

She heard Scobee's voice in her headset. "Fifteen."

At -00:00:10, Scott McAuliffe's third-grade classmates in the bleachers began chanting the countdown. "Ten, nine, eight . . ." At -00:00:06, the orbiter's main engines fired. The crew felt three booms from below that shook their seats, the floors, and the walls. The engines' force made the 4.5-million-pound stack wobble with the twang effect that shifted the stack sideways by two feet, a lurch the Challenger Seven felt as the stack flexed back to upright. "Three at a hundred," Scobee said. All three engines were at full power, burning liquid oxygen and hydrogen at a temperature of 6,000 degrees, hot enough to boil steel.

At 00:00:00, the solid boosters fired. The explosive bolts holding the stack to the launchpad detonated, releasing *Challenger* from its shackles. Liftoff was officially recorded at one one-hundredth of a second after 11:38 a.m. Eastern Standard Time on January 28, 1986.

The stack rose slowly in fire and steam, burning ten tons of fuel per second as Scobee told his crew, "Here we go."

—◆—

Hugh Harris's voice came over the loudspeakers at KSC: "Liftoff! Liftoff of the twenty-fifth space shuttle mission, and it has cleared the tower."

The roar of liftoff shook the VIP bleachers. Cheers went up from there to the roof of the Launch Control Center, the press center, and the roads leading to the Cape, where space fans watched the launch from their cars and trucks. The Teacher in Space semifinalists were back in the bleachers, missing yet another schoolday, looking forward to their upcoming year as NASA ambassadors in schoolrooms all over the country. Barbara Morgan, narrating the launch for a NASA TV crew, whooped and applauded. "Bye, Christa. Bye-bye, Christa!" she said.

At liftoff the minus sign on the countdown clock became a plus sign. At +00:00:07, jurisdiction over *Challenger* shifted to Mission Control, in Houston. Now it was JSC public affairs officer Stephen Nesbitt's voice on TV broadcasts and the loudspeakers at the Cape. "Challenger now heading downrange," he said. "Three engines running normally, velocity 2,257 feet per second, altitude four-point-three nautical miles, downrange distance three nautical miles. Engines throttling up."

—◆—

On the flight deck, the normally tight-lipped Mike Smith eyed his instrument panels while the stack gained speed. "Go, you mother," he said.

Resnik added a comment NASA would redact in transcripts of the flight log. "Shit hot!"

Scobee gave that an audible shrug. "Oooh-kay . . ."

As the stack climbed, its momentum pressed them to their seats with three times the force of gravity. Scobee liked to compare the sensation of launch to riding "a runaway train. There are metallic bangs and clanks, and the shuttle vibrates and shakes." According to Hoot Gibson, "You can't hear yourself think. Your noise sense is totally maxed out—it's as noisy as it can possibly be. You are slapped back in your seat, and that launch tower is gone. You're feeling the shock waves off the engines hitting the ground, reflecting back and just bathing the vehicle in vibration." Sally Ride called her first launch "emotionally and psychologically overwhelming. The actual experience is not even close to the simulators! You are sitting on top of tons of rocket fuel and it's basically exploding underneath you. It's exhilarating, terrifying, overwhelming."

At +00:00:16, the stack arced over the Atlantic, leaning backward for an optimal trajectory toward orbit. For a moment, rather than rising at a steep angle as if on a roller coaster, the crew was upside-down. The engines cut back to 65 percent power, per the mission plan, as the stack entered a high-altitude zone where they could expect turbulence. "Hold on," Scobee told his crew.

The next half-minute held more bone-rattling jolts and noise than anything they'd felt in training. "Ten thousand feet," Smith reported. Entering the stratosphere, the orbiter shook like a car doing a hundred on a rutted country road. Then, at +00:00:40, it broke the sound barrier. "There's Mach One," Smith said. Accelerating to two thousand miles per hour at fifty thousand feet, above the rough air, they were free to go faster. At +00:01:07, Mission Control radioed the good news: "*Challenger*, go at throttle up."

Scobee said, "Roger, go at throttle up."

They were rising at twice the speed of sound when Smith saw a flash of vapor and fire outside the forward windows. Or did he see one of the rocket boosters too close to his window? Whichever it was, Christa would have heard the surprise in Smith's voice on her comm line. He said, "Uh-oh."

12

A MINUTE AND TWELVE SECONDS EARLIER, A SEAL IN THE rocket booster on the shuttle's right flank had let out a puff of smoke. Nobody saw it at the time, but that bit of smoke was the first sign of trouble. It came from a barely measurable breach in one of the rubber seals called O-rings that held the booster together. The cold had stiffened the rings enough to keep one of them from expanding to fill the seal between booster sections, a gap that measured less than a millimeter, not much wider than a human hair. The breach allowed a few grams of superheated fuel to burn through. Several more puffs of smoke followed the first—all subsumed in clouds of white exhaust as *Challenger* rose off the pad.

Smoke from an O-ring was a signal of disaster. In most cases it would lead to a chain reaction in which the torpedo-shaped booster, packed with rocket fuel the texture of a pencil eraser, would ignite all at once. The shuttle would blow up before it cleared the tower. In this instance, due to a momentary miracle NASA engineers and a presidential commission would puzzle over for four months, a thumbnail-sized chunk of solid fuel inside the booster broke off and plugged the gap in the seal. Instead of exploding, the stack kept rising off the launchpad.

The crew's immediate families watched from the roof of the Launch Control Center a safe three and a half miles away. The roof shook under their feet as the shuttle cleared the tower. Steve McAuliffe clutched his children. June Scobee shouted "Go, baby, go," cheering along with Jane Smith, her son, and her two little girls; Cheryl McNair, with a toddler, Reggie, and baby daughter, Joy; Lorna Onizuka, with her two daughters, Janelle and Darien; Bruce Jarvis, Greg's father; and Marvin Resnik. They all shielded their eyes as the shuttle climbed. They heard the NASA TV feed over loudspeakers, the same commentary millions of schoolchildren heard. At +00:00:07, they heard Steve Nesbitt, the voice of Mission Control in Houston, narrating what seemed to be a routine ascent. "Challenger now heading downrange . . ."

At +00:00:37, the accelerating shuttle encountered the first of several wind shears. "A lot of wind here today," Smith said on his intercom. Still, the shuttle performed perfectly. While Scobee, Smith, and Resnik kept an eye on the 1,300 switches and dials on the flight deck, their onboard computers kept them on course. Then their luck changed. The strongest wind shears ever recorded during a shuttle mission jarred the stack enough to dislodge the chunk of fuel plugging the leak in the right-hand rocket booster.

The first sign of danger at Mission Control was a line of numbers on the flight controllers' low-resolution computer screens: a falling pressure reading from the right-hand rocket booster. The booster was leaking fuel.

A flicker of flame appeared at the edge of the faulty O-ring. At first it was the size of the flame on a cigarette lighter, but it grew larger and hotter. Drawn downward and focused by the shuttle's slipstream, the fuel leak acted like a blowtorch on the surface of the external tank. The flame took only three seconds to penetrate the tank's aluminum skin.

At +00:01:12 and seventy-seven hundredths of a second, the external tank ruptured, spilling tons of liquid hydrogen into open air eight miles over the Florida coast. The condensed fuel turned instantly to gas—the term is "subliming"—the same flammable hydrogen that had filled the *Hindenburg* half a century before. On contact with the flame from the booster the gas ignited. Smith, sitting on the right-hand side of the flight deck, would have been first to see the fire that quickly engulfed the orbiter, spewing white vapor in all directions. This is the moment when he said, "Uh-oh."

At +00:01:13 and fourteen hundredths of a second, less than a heartbeat after the explosion, what was left of the leaking booster swiveled sideways on its broken strut and smashed into what was left of the external tank. TV viewers saw the rocket boosters' white exhaust as the untethered boosters went in different directions, crisscrossing in flight.

What viewers didn't see was what became of the orbiter. Torn from the rest of the stack, it seemed to disappear in the explosion.

While television viewers saw the stack burn up, the view was different from the roof of the Launch Control Center and the bleachers at Cape Canaveral. From there, the shuttle was a speck in the sky more than thirteen miles away, barely visible through binoculars. Billows of exhaust from the boosters might be part of the mission plan. "We didn't know right away. We had never seen a liftoff," Christa's sister Lisa remembered. Even astronaut Rhea Seddon saw the smoke in the sky as a good sign at first. "Look," she said, "you can see the boosters coming off." Another astronaut nearby said, "No, it's too early."

The rocket boosters kept flying upward on diverging paths, their exhaust trails forming a pitchfork shape in the sky. One of them did a 360-degree flip in midair. Air force major Gerald Bieringer, a little-known NASA functionary whose title was range safety officer, tracked the still-burning boosters toward New Smyrna Beach, fifty miles north of Cape Canaveral. If they crashed before expending their fuel they could destroy ships, houses, city blocks. Bieringer, sitting in a military bunker at the Cape Canaveral Air Force Station, activated switches that armed and fired the boosters' self-destruct packages, possibly saving dozens of lives. The boosters exploded and fell into the sea.

By now Christa's family knew something was terribly wrong. "We could hear people crying and screaming," her sister said.

June Scobee, watching from the roof of the Launch Control Center, "prayed for a miracle. I looked at Steve McAuliffe. Our eyes met." Steve knew. So did Mike Smith's daughters. One of them cried, "Daddy, I want you! You promised nothing would happen!"

Computer screens at Mission Control in Houston and Launch Control in Florida showed white *S's* instead of the usual rows of numbers. The *S* stood for static. It meant there was no telemetry from *Challenger*, no audio, nothing.

"Obviously a major malfunction," Nesbitt announced.

Pieces of the stack began falling to earth. Some left contrails, leading a few spectators to think the crew was parachuting to safety. "They're coming back!" someone shouted.

Bob Veilleux, the New Hampshire science teacher who was there as one of Christa's runners-up, knew the shuttle had no parachutes. He recalls "a minute, it might have been two or three, when some of us knew they were gone." The rest would know soon enough. At +00:02:43, Nesbitt's voice came over the loudspeakers: "We have a report from the flight dynamics officer that the vehicle has exploded."

Photographers at the Cape closed in on Christa's parents. Ed Corrigan hugged grief-stricken Grace while clutching their daughter Lisa's hand. Before long, a NASA executive named Robert Brown approached them, saying he was sorry. "The craft has exploded." He led the Corrigans from the bleachers to a waiting bus while Grace repeated, "The craft has exploded. The craft has exploded." The bus delivered Christa's parents to the astronauts' quarters in the O&C Building, where Brown asked them to wait for Vice President Bush. Bush had hurried from the White House to board Air Force Two and fly to Florida to meet the families.

"We found our son-in-law Steve McAuliffe and grandchildren Scott and Caroline in Christa's dormitory room," Grace remembered. "Christa had been there only that morning. Her sneakers were on the floor. . . . Steve's first words were, 'This is not how it's supposed to be!'"

Flight operations director George Abbey appeared at the dorm-room door. Abbey looked stricken. He shook Steve's hand without a word, then turned and trudged down the hall.

—◆—

The mood at Concord High School had been festive all morning. Principal Foley and football coach LeBrun patrolled the halls on what everyone expected to be the best day in school history. In the cafeteria, students and teachers wearing Christa buttons and party hats watched the telecast on a TV mounted on an AV cart. Reporters kept asking how they were feeling. A pair of teen rebels bucked the happy tide, waving a hand-lettered sign: I'D RATHER BE LEARNING. "All we hear is Christa this and Christa that," one of them said. Which didn't keep them from watching the countdown.

"I remember the explosion, the two streams of white smoke," another student said, "and realizing there was no shuttle in the middle. I remember one of our teachers standing up on a cafeteria table and shouting, 'Everybody shut up! Shut the hell up, something's wrong.'"

Math teacher Susan Capano would remember hearing Nesbitt's voice on the TV: "The vehicle has exploded." At that, "one of the girls in my classroom said, 'What do they mean by "vehicle"?' I said, 'I think they mean the shuttle,' and she got very upset with me. She said, 'No! No! No! They don't mean the shuttle!'"

Principal Foley told students and teachers to go to their classrooms. He ordered reporters and photographers to "Get out. Now." When several newsmen ignored him, LeBrun's football players took them by the elbows and led them out of the building. One writer stopped Eileen O'Hara on the way to her car. She had talked to Christa on the phone the night before, when Christa said she was eager to fly after so much waiting. "I think Christa knew the risks, and she took them," O'Hara said through tears. "She encouraged kids to try to go beyond anything they've done before, and if things don't work, you should still try."

A librarian hurried from the high school to a nearby church, where she rang the bells seven times, once for each member of the crew. While the bells rang, a police car pulled over outside the church. A policeman got out and knelt in the snow.

—◆—

President Reagan had been reading a draft of his State of the Union speech when Bush and National Security Advisor John Poindexter rushed into the Oval Office. Bush said, "There's been a serious incident with the shuttle."

Reagan said, "Is that the one the teacher was on?"

As the news spread in Washington, Congress shut down for

the day. The president went through with a noon meeting with network news anchors. His remarks were off the record, but an attendee took notes.

"It's a horrible thing that all of us have witnessed and actually seen it take place," Reagan said. "I'm quite sure also, when you look at the safety measures that sometimes those of us looking on have gotten impatient with when flights have been aborted . . . it seemed as if they were taking things too seriously. Now we know they weren't."

"Mister President," he was asked, "sending civilians into space was based on the assumption that it was routine to go into space, that it was now safe—even a teacher we could send up. Do you think that notion is now gone?"

"Well, what could you say? Other than that, here was a program that had a hundred-percent safety record."

At that point, Patrick Buchanan, Reagan's communications director, broke in. "One more question," Buchanan told the TV anchors.

Asked if he had any words for schoolchildren who had watched the explosion, Reagan said, "The world is a hazardous place, always has been. In pioneering we've always known that there are pioneers who give their lives out there on the frontier. And now this has happened. It probably is more of a shock to all of us because of the fact that we see it happen—thanks to the media—not just hearing about it as if something happened miles away."

One of the newsmen asked if he had any "special thoughts" about Christa McAuliffe.

"I can't get out of my mind her husband and her children. But then that's true of the families of the others." At the same time, the president said, "The families of the others had been a part of this whole program and knew that they were in a hazardous occupation." As for Christa's family, "Knowing that they were there and watching . . . well, your heart goes out to them."

———◆———

NASA officials led Christa's family into an auditorium in the O&C Building. "We were not given any information," Grace Corrigan recalled, "except that the Vice President was on his way." She went to the dining room where Christa and the rest of the crew had picked at their steak-and-eggs breakfast that morning. The vanilla cake iced with the mission logo—the cake they'd promised to eat when they got back to Earth—was still in a refrigerator. Grace returned to the auditorium with "coffee, tea, and cold drinks that nobody wanted."

Several family members still held out hope. "Very, very faint hope," Grace called it. Then George Abbey, the flight operations director who chose astronauts and assigned them to their missions, asked for the families' attention. "All the crew members are dead," he said. "They could not have survived." Greg Jarvis's father swooned and was treated for shock. Lorna Onizuka, leaning against a wall near a panel of light switches, slid to the floor, turning the room dark for a few seconds. When the lights came back on, NASA doctors approached Christa's father. "My husband was horribly angry," Grace Corrigan recalled, "so angry I thought he was going to have a heart attack. They tried to give him medication. He refused." In a control room nearby, Elmer Thomas, a sixty-nine-year-old engineer who had helped fuel the external tank that morning, slumped over his desk with a fatal heart attack.

In the auditorium, Cheryl McNair hugged Ron's brother Carl. "What am I going to do now?" she asked.

June Scobee wondered the same thing. "It was a nightmare

we'd never wake up from. I had no husband and no answers." Later she found her husband's briefcase in his dorm quarters. It held his wallet and keys, flight manuals, star charts, family photos, and a Valentine card that read, *For My Wife*. He had been thinking ahead as usual: *Challenger* was scheduled to fly for six days and land the week before Valentine's Day.

June spoke with Steve McAuliffe and the other family members while they waited for the vice president. She convinced them to join her in a unified message to Bush and the Reagan administration.

The sun was setting when Air Force Two touched down at the Cape. Joined by spacefaring senator John Glenn, the vice president met privately with the *Challenger* families, assuring them that the nation shared their grief. Glenn gave a brief, heartfelt talk. At sixty-four, the senator from Ohio was almost as lean and intense as he had been as a Mercury astronaut a quarter century before. "There are times . . ." he began. Fighting his emotions, Glenn began again. "There are times when you devote yourself to a higher cause. I know that seven brave heroes were carrying our dreams and hopes with them today. We will carry their memories with us."

The meeting was ending when June Scobee asked to say a few words on behalf of the families.

"We've been talking," she told Bush and Glenn. "My husband . . ." June choked up, then went on. "My husband believed in the exploration of space. From talking to the other families this afternoon, I know that was true of everyone on that shuttle. And if we let this setback halt the space program, then those seven people died for no purpose. That's not what Dick would have wanted. None of them would have wanted that. So we want you to make the space program better than ever." She looked straight at the vice president. "Make their deaths count for something."

Bush's plane was waiting. He was on his way out, led by Secret Service agents, when he called June aside.

"I know loss," Bush said. "Barbara and I lost a daughter." Pauline Robinson Bush had died of leukemia in 1953, at the age of three. Now he handed June a slip of paper with his home phone number. "Call me anytime."

———◦◦◦———

The countdown clock ticked for hours after there was no one in the bleachers to watch it. It clicked through +02:00:00 and the three- and four-hour marks. Finally, a pair of NASA sedans pulled up at the flagpole beside the clock. The cars' drivers got out and lowered the flag to half-staff. The clock went dark.

———◦◦◦———

TV coverage of the disaster ran uninterrupted as ABC, CBS, and NBC each devoted hours of programming to the *Challenger* disaster. Viewers saw the explosion in a seemingly endless loop. CNN's nonstop coverage would help make the all-news channel a rival to the three broadcast networks when news was breaking.

NBC's Tom Brokaw called the accident "a nightmare, a cruel, shocking end to what everyone expected to be a triumph." Peter Jennings opened ABC's *World News Tonight* this way: "It is the worst disaster in the history of the American space program. President Reagan has declared a week of mourning for the seven astronauts, five men and two women, who lost their lives on their way into space this morning." Jennings called the disaster "an enormous shock to the manned space program . . . never before have American lives been lost in space." CBS anchor Dan Rather struck a poetic note, quoting a sailors' prayer: "Thy sea is so great, and my boat so small."

Reagan postponed his State of the Union address. Instead, he would to speak to the nation about *Challenger* that night. Speech-writer Peggy Noonan had less than three hours to prepare a draft of his remarks. "The president has to speak to the children and reassure them that the world isn't ending," Reagan's advisor Dick Darman told her. Noonan watched replays of the stack's fiery breakup, rewinding the video to the astronaut walkout that morning, when Scobee's smiling, waving crew boarded the Astrovan.

Reagan's televised speech that night would be remembered as one of the best of his presidency.

"Ladies and gentlemen," he began, "I'd planned to speak to you tonight to report on the state of the union, but the events of earlier today have led me to change those plans. Today is a day for mourning and remembering. Nancy and I are pained to the core by the tragedy of the shuttle *Challenger*. We know we share this pain with all of the people of our country." He alluded to the national yawn that had greeted recent shuttle flights. "Perhaps we've forgotten the courage it took for the crew of the shuttle . . . the Challenger Seven were aware of the dangers, but overcame them and did their jobs brilliantly." Reagan named them. "We mourn seven heroes: Michael Smith, Dick Scobee, Judith Resnik, Ronald McNair, Ellison Onizuka, Gregory Jarvis, and Christa McAuliffe. We mourn their loss as a nation together." Then he addressed "the schoolchildren of America who were watching the live coverage of the shuttle's takeoff." The president's words were by all accounts a true expression of his feelings, channeled by Noonan. "I know it is hard to understand," he said, "but sometimes painful things like this happen. It's all part of the process of exploration and discovery. It's all part of taking a chance and expanding man's horizons. The future doesn't belong to the fainthearted. It belongs to the brave."

In closing, Reagan quoted the poem every flyer knew. "We will never forget them," he said, "nor the last time we saw them, this morning, as they prepared for their journey and waved good-bye and 'slipped the surly bonds of Earth' to 'touch the face of God.'"

Reagan's address to the nation set the tone for all that came later. *People* magazine put a smiling Christa on its cover, hailing the teacher who "set out to reawaken the pioneer spirit in Americans," as the magazine put it, until she and the others "died a little more than a minute into their mission." NASA and Morton Thiokol redesigned the space shuttles' rocket boosters to make them safer. The agency paid Rockwell $1.7 billion to build a new shuttle, *Endeavour*, to replace *Challenger*. The full story had yet to be told.

13

IN THE FIRST MINUTES AFTER THE EXPLOSION, WHEN THE WORLD realized that Christa and the others were lost, they were still alive. *Challenger*'s welded-aluminum crew cabin survived the explosion intact. All seven men and women aboard were conscious, at least at first, and aware that something was wrong.

They were eight miles high and rising fast when Scobee radioed his last transmission to Houston: "Roger, go at throttle up." The next three seconds wrapped them in flames that lit up millions of TV screens. Viewers believed that nobody could survive such a conflagration. But the crew did not burn up. The shuttle's skin of heat-resistant silicon tiles, made to withstand the hellfire of reentry, protected them. There was no fire inside the crew cabin. The temperature inside may have risen by several degrees but no more. At T +00:01:13, the crew heard Mike Smith say "Uh-oh" as the shuttle got an unexpected boost: two million pounds of sudden thrust from the explosion of the external tank, sending the orbiter barreling upward even faster. NASA's dry account placed "the start of total vehicle breakup at 73.213 seconds after launch . . . the released fluids vaporized rapidly. The last telemetry from *Challenger* was received 73.618 seconds after launch." The last signal from the shuttle was a crackle of static on the air-to-ground audio circuit, indicating that Mission Control had lost contact with the shuttle.

Christa and the others would have felt their sudden burst of speed as a downward push. "You feel like you're melting into yourself," she'd said of similar G forces during training, "like the Wicked Witch of the West." She could only hold on for dear life while the crew cabin, propelled by its own momentum and the blast of the tank's explosion, kept going up.

At first, the flight-deck crew had a view of blue sky beyond the cataclysm. Scobee, Smith, Resnik, and Onizuka may have seen white vapor and chunks of external tank fly past their windows. Then their headsets went dead. They may have seen one or both rocket boosters as the boosters broke free and corkscrewed upward on their own.

According to the presidential commission that would investigate the accident, "The orbiter, under severe aerodynamic loads, broke into several large sections which emerged from the fireball." *Challenger*'s left wing broke off. Its tail section tore loose, engines still burning, and began falling toward the ocean. "The forward fuselage, trailing a mass of umbilical lines, pulled loose from the payload bay." Those lines had fed air and power to the crew cabin, which went dark; now they trailed uselessly behind it.

The crew cabin lurched to the right at first and then twanged back, rattling like a runaway train. On the dim middeck, Christa, Jarvis, and McNair had no dials or gauges to give them an idea of what was wrong. Their experience of the explosion's aftermath began with deafening noise and G forces that pressed them downward with suffocating force. Their training had accustomed them to three Gs, enough to make Christa feel as if she weighed three times her 128 pounds. This was worse. Enough to hurt. The sudden acceleration drove them into their creaking steel seats with a force of twelve to twenty Gs, enough to make Christa feel she weighed a ton and to buckle the bolts that held her seat to the floor. She could not lift a finger.

Astronaut Joseph Kerwin, an MD who led NASA's initial investigation, would describe such a jump in speed and G forces as uncomfortable but "survivable, and the probability of injury is low."

"The noise had to be deafening," says Frank Hughes, the crew's flight-training chief. "Everything's shaking. But the G forces are survivable, and these are brave people. They would fight to stay alive."

After a moment of crushing acceleration, the pressure eased. The crew cabin was still going up from 46,000 feet—its altitude when the stack broke apart—but slowing as it rose. G forces dropped to about four Gs—uncomfortable but not painful.

"What would they do then? Scobee and Smith would try to fly home," says Kerry Joels, a former NASA scientist and a curator of the Smithsonian's National Air and Space Museum. "It wasn't a question of piloting—Dick Scobee and Mike Smith could fly that thing blindfolded and bring it down on a dime. They knew all the switches inside out." Scobee may have flashed on practice runs including the simulator snafu that left him making a perfect landing in total darkness. This might feel like a simulation—that was the idea behind all their training, to prep them for anything—but here was the ultimate set of green cards: *ET ignites. No comm. No power.* "It's not even a matter of 'Is this survivable?' They would have had a sense that the shuttle wasn't airworthy," Joels says. "They would have thought it might be impossible to fly after it broke apart, but that wouldn't keep them from trying."

Each of the shuttle's ten windows consisted of three panes of heat- and shock-resistant glass, with a middle pane that was more than an inch thick. A broken window or blown hatch would trigger a quick depressurization in which air whooshed out of the crew cabin, but NASA's forensics would show no sign of a sudden, catastrophic loss of pressure. The windows held. Still they were losing pressure and air, possibly through holes caused by shrapnel thrown off by the explosion, certainly through breaches in the aft section of the crew cabin where it had torn away from the rest of the shuttle.

They still had electrical power in the first seconds after the explosion. That would give Scobee time to thumb the red button on the control stick by his knee as the crew cabin climbed. As another astronaut put it later, "They knew the situation was perilous, but they were in a cockpit with a control stick and there was a runway only twenty miles away. They believed they had a chance."

In all likelihood they kept their visors down. That would help protect them from a sudden depressurization. Better to breathe the oxygen that was left inside their helmets than lift their visors and let it escape.

According to six-time shuttle astronaut Jim Wetherbee, the only American ever to command six space missions, "If I were in Scobee's seat, the first thing I would do is take a moment to assess what was happening." Flyers and doctors are trained to follow the same rule: *First, do no harm.* Do nothing rash that might make matters worse. At the same time, Scobee would have known he had no time to lose. Next, Wetherbee says, "I would go to panel F4," a set of buttons on the commander's side of the flight deck.

In the instants after the explosion, while they still had electrical power, "the AUTO buttons beside the PITCH and ROLL/YAW buttons would be illuminated. The craft was on AUTO," meaning that it was controlled by computer, as usual during a shuttle's ascent to orbit. "To try to take control, he would push the two CSS buttons to the right of the AUTO buttons—CSS for Control Stick Steering." Pressing those buttons would allow Scobee "to take the controls himself, and try to fly." If they worked. "That's what you would do in a sim. That's how a commander thinks."

Scobee would have tried to radio Mission Control. "Houston, this is *Challenger*." But the air-to-ground circuit was dead. The CSS button would not work. Seconds after the explosion, their intercom was dead, but Scobee and Smith had been through hundreds of real and simulated emergencies during their flying careers and astronaut training, and they didn't need to discuss what to do. According to astronaut Mullane, a former air force flyer, "hand signals would have worked perfectly well."

Using hand signals, the commander or pilot would indicate a panel on the console between them. By now the flight deck's computer screens had gone dark. Every button that should have been lit had gone dark. The only light in the shuttle was sunlight from the windows. Another last-ditch option was toggling a pair of protected dials labeled SRB SEPARATION and ET SEPARATION from the computer-controlled automatic setting, marked AUTO, to MAN for manual. This is one way commanders and pilots were trained to react to uphill emergencies in a simulator. The idea was to ditch the rockets and external tank in hopes of flying the orbiter back to the landing strip at the Cape. Once the switches were toggled upward from AUTO to MAN, lifting a pair of plastic shields and pressing the buttons marked SEP would jettison any remaining part of the solid rocket boosters or external tank. If only the buttons would work.

They didn't; nothing happened.

Scobee's side of the flight deck held an ABORT button beside a dial marked ABORT MODE. "That's going to be their goal," says Joels. The abort mode dial had five settings: There was OFF, the default setting; plus RTLS, which stood for "return to launch site"; TAL for "trans-Atlantic," designed to redirect the shuttle to a NASA runway in Africa; AOA for "abort once around," which would to return them to Cape Canaveral after one orbit of the earth; and ATO for "abort to orbit," a stopgap designed to leave them circling the planet while Mission Control figured out what to do. The latter three settings were not viable options for a ship with no power. That left RTLS. "They knew the engines were gone. The only thing to do is try to glide home," Joels says.

Some of the flight-deck switches were protected to keep them from being thrown by accident. To arm them, a crew member had to lift a spring-loaded plastic shield. Only then could he or she throw the switch or turn the dial beneath. NASA's investigators would confirm that at least some of the protected switches on the flight deck were thrown after the explosion.

The astronauts had emergency air packs—seat-mounted canisters of breathable air connected to their helmets by thin tubes. These personal egress air packs were called PEAPs, pronounced "peeps." They were intended for the sort of launch-pad evacuation Resnik and Hank Hartsfield had contemplated before their *Discovery* launch two years before. PEAPs could be crucial to surviving a mishap on the pad, but could not keep an astronaut alive for long. Each held five to six minutes of breathable air. And due to design considerations and compromises, the commander's and pilot's PEAPs were mounted behind their seats. Scobee and Smith couldn't reach them without unstrapping and clambering behind their seats to switch on their air. That was no option now. They would need one of the mission specialists seated behind them—below them in the shuttle's uphill orientation—to do it for them.

That is what happened. NASA's forensics would show that at least three of the crew's PEAPs were activated in the moments after the explosion. Resnik, seated directly behind Scobee and Smith, had been trained to activate the commander's PEAP in an emergency. Onizuka, seated to her right, had been trained to activate Smith's. Like throwing a protected switch, this called for two distinct motions: pulling a lever, then twisting the lever 180 degrees. Such a move could not happen accidentally. It could not be caused by the buffeting or G forces the shuttle endured. NASA's investigations would show that Smith's PEAP had been activated—probably by Onizuka, who could reach the pilot's PEAP from his seat behind Smith's.

Astronaut Mullane would ask himself how he would have reacted. "I wondered if I would have had the presence of mind to do the same thing. Or would I have been locked in a catatonic paralysis of fear?" There were no green cards for in-flight PEAP activation. "The fact that Judy or El had done so for Mike Smith makes them heroic in my mind. They had been able to block out the terrifying sights and sounds and motions of *Challenger*'s destruction and had reached for that switch. It was the type of thing a true astronaut would do—maintain their cool in the direst of circumstances."

There is evidence that Smith tried to restore power to the shuttle. A panel on his side of the flight deck held a set of flat paddle switches and round toggle switches under the label FC/MAIN BUS. The pilot could throw paddle switches with one motion, but the toggles, like the PEAP dial that was giving him air to breathe, took two motions. He had to lift and then turn those toggle switches in an attempt to activate them. NASA's investigators would find some of those switches thrown, which could not have happened by accident.

At T +00:01:23, ten seconds after the explosion, *Challenger* was eleven miles high and still rising. "All the lights have gone off. The cockpit is dead," Wetherbee says. "The vehicle's dying." Still they tried to stay alive. In the direst estimate the crew had no more than ten seconds of consciousness left—assuming a quick depressurization, a painful spell of ringing ears, and the excruciating, burning feeling that their blood was boiling as the cabin depressurized. But it is just as likely that they had anywhere from half a minute to more than two minutes of awareness before the crew cabin crashed into the Atlantic. Mullane pictured Christa, Jarvis, and McNair on the middeck, "locked in the most horrifying of circumstances. They had no windows, no instruments. They were totally dependent upon the upstairs crewmembers to keep them informed."

The crew cabin rose for twenty-five seconds on its own inertia and the force of the explosion, slowing from thousands of miles per hour to hundreds and then to something closer to a walking pace until its momentum ran out at an altitude of sixty-five thousand feet. Mission time was T +00:01:38. To the crew the last instants of *Challenger*'s ascent would feel like reaching the top of one of the Vomit Comet's roller-coaster parabolas: an instant suspended in space before the bottom drops out. In scientific terms they were at the apogee of a ballistic arc. Like a bullet shot upward, the crew cabin had risen from a height of 8.7 miles at the moment of the external tank's explosion to 12.3 miles, where gravity finally overcame the crew cabin's momentum and they started to fall.

At sixty-five thousand feet the stratosphere thins to the point that the sky begins to turn from blue to black. At twenty minutes to noon Florida time, *Challenger*'s flight-deck windows would have showed the first few stars in a darkening sky.

The panoramic view in the flight deck's forward and overhead windows would have showed sky at first, then horizon as the shuttle reached its highest point and began to turn nose-down. After that, their view would be nothing but ocean.

Resnik used to say that the difference between training and flying is that training is harder. Most flights followed the mission plan like clockwork, but simulations never did, not with sim sups handing you two or three green cards at a time. That's why many astronauts considered prelaunch sims the most difficult part of their training. Resnik called them "challenging, but I wouldn't consider them difficult." By the time a shuttle left the launchpad and they took up their flight duties, "it's just like you practiced it."

Not this time. They were in a situation no NASA crew had ever faced.

The first studies of the accident would suggest that the crew could not have stayed conscious on the way down. "They would have tried to take control of the vehicle, because that's who they *are*, but this is unsurvivable," a NASA scientist would say. "The cabin is tumbling and spinning like a top. They got thrown in every direction. Their seats are coming unbolted. You've got seats and people flying around the cabin."

That view would prove to be guesswork. In fact, what was left of the orbiter fell from its twelve-mile height without tumbling or spinning enough to kill or injure the crew, arrowing downward nose-first due to an aerodynamic effect no one could have anticipated. As the cabin fell, the torn umbilical cables and wires that had connected it to the payload bay steadied its flight.

John Young watched it happen. Young, the fifty-five-year-old head of the Astronaut Office, had walked on the moon in 1972.

He knew and liked Scobee and all six members of Scobee's crew. Young had piloted the two-seat T-38 that flew McNair from Houston to Cape Canaveral five days before the launch. He was following *Challenger*'s ascent from the cockpit of one of the planes NASA used to observe shuttle launches, and he heard reports of an aerodynamic effect he didn't expect but immediately understood. "We found out later that it trimmed in flight, nose-down," he said. By "trim" he meant that the plummeting crew cabin found the path of least aerodynamic resistance. With its severed wires and cables trailing behind like the tail of a kite, the cabin stabilized and fell smoothly.

Launch director Gene Thomas, sitting in a hushed Firing Room at the Launch Control Center, watched the trails of white smoke in the sky. As he recalled, "I sat in shock and prayed to God that through some miracle the crew could be saved." Scobee had always reminded him of Hollywood cowboy John Wayne. "My first hope was to see *Challenger* sail clear of the fireball and slowly circle to begin a controlled glide back to safety at the KSC landing facility. Ace flyer Dick Scobee, good old John Wayne, had taken manual control and would bring *Challenger* home." At least that's what Thomas hoped.

"If Dick Scobee and Mike Smith tried to fly what was left of their vehicle, more power to them," says Wetherbee, "but there is no way out of this. At some point I would have said, 'We're screwed,' and tried to make peace with my maker."

"Scobee fought for any and every edge to survive," said astronaut Bob Overmyer. "He flew that ship without wings all the way down."

"If it had landed softly," an investigator said, "they could have swum home."

"A strong person on the middeck, like Ron McNair, might have escaped by bailing out," Young said.

McNair would not have bailed out without helping Christa and Jarvis get out the hatch first. In any case, they had no parachutes. If McNair looked out his porthole in the hatch he saw sky and sea.

A devout Baptist, he had often told his brother Carl why he felt safe riding a rocket into orbit: "I wouldn't set foot on that shuttle if Jesus wasn't there first." Ron McNair also loved the barely harnessed power it took for a shuttle to escape Earth's gravity. He used to bring a videotape of his 1984 mission when he visited his brother. "He turned the volume up to full power, shoved the bass over to max, and sat entranced as the TV screen replayed the explosive ignition," Carl recalled. "My living room floorboards rumbled with the thunder of the rocket engines. Eventually he blew out my brand-new speakers."

As *Challenger* fell there was little to hear. Bolts creaked, wind whistled as they nose-dived toward the Atlantic. When Carl pictured his brother sitting by the hatch on the middeck, he recalled Ron's sonorous voice during church services. He reminded himself that the shuttle's descent took "long enough to recite the Lord's Prayer."

Trimming downward with its torn umbilicals trailing behind, the crew cabin reached a terminal velocity of more than two hundred miles per hour, the speed of a peregrine falcon in full dive, yet it had so far to fall that its final descent took more than two minutes.

Before the flight, Greg Jarvis had told his mother not to worry. "We know we're sitting on a keg of dynamite, but they have this thing down to a routine," he said.

Now Christa sat beside him, comms dead.

"One of the hardest things for a teacher to do," she once said, "is to experience something and not be able to tell anybody about it."

Her father would picture her buckled into her seat, next to Jarvis. "I'd lay a bet that when they went down, Greg and Christa were holding hands," Ed Corrigan said.

—◆—

After the external tank blew up, the crowds at the Cape heard Nesbitt's dry commentary: "Obviously a major malfunction. We have no downlink. We have a report from the flight dynamics officer that the vehicle has exploded."

The crew was alive. At that moment the crew cabin was nineteen seconds into its dive toward the sea off Cape Canaveral.

In the bleachers, the Corrigans clutched each other.

Monitor reporter Jimenez ran for the press center.

June Scobee stumbled on her way down the stairs from the roof of the Launch Control Center, feeling she was in "a nightmare that you try to piece together and then realize it isn't rational. My head knew they were dead, but my heart did not."

There was so much telephone traffic that phones went dead all over the Cape.

Mullane, watching one of the first replays of the explosion, saw *Challenger* appear to "disintegrate into an orange-and-white ball. . . . Streamers of smoke arced toward the sea." The crew cabin was one of those streamers.

Jeff Hanley, a college student working part-time for NASA, was on the roof of the Launch Control Center with the crew's immediate families. "A crystal-clear morning, but cold," he recalled. "Then the explosion happens. Tremendous confusion on the roof. What are we seeing? What are we watching? 'Obviously a major malfunction' over the loudspeaker. We're all looking. We're looking, looking, looking. When is the shuttle going to emerge and head back toward the landing strip right there on the property? We keep looking, keep looking. Of course it's nowhere to be seen. Then off in the distance, after a few moments, you see a sparkling in the sky below the airburst."

They were seeing the sun's reflection off hundreds of pieces of the stack, including the crew cabin. "A sparkling effect," Hanley called it, "as the debris was falling into the ocean."

14

MILLIONS OF TELEVISION VIEWERS SAW *CHALLENGER* APPEAR TO blow up, leaving a pitchfork of white exhaust in the sky.

NASA executives, flight controllers, engineers, and technicians watching from the Cape and at Mission Control in Houston tried to figure out what they were looking at. The flight controllers were most concerned with the crew cabin in the orbiter's nose, but there was no sign of it in NASA's telemetry, which was locked down seconds after the explosion. "All the doors in the Launch Control Center were immediately locked, the telephone lines all disconnected, and all data associated with the flight frozen on the screen and immediately impounded," one of the engineers at KSC recalled. *Challenger*'s data was now a government secret, accessible only to management.

There was no sign of the crew cabin in the waters off the Cape, where charred pieces of fuselage and external tank began splashing down within minutes of the explosion. NASA sent helicopters offshore, warning the pilots to "stay clear of the falling debris."

Some pieces of *Challenger* sank; others floated long enough to wash up on the beaches of Florida's Space Coast in the hours and days that followed. (Even today, thirty-five years later, some still do.) But none came from the crew cabin.

Yet George Abbey was surely right. "They could not have survived." Without ejection seats or any other means of escape, Christa and the others had no way out.

"My only comfort was my belief that their deaths had been mercifully quick," astronaut Mullane wrote in his memoir, "the instantaneous death we all hope for when our time comes. There were more than a million pounds of propellant in the tank when it detonated. The explosion must have destroyed the cockpit and everything in it."

Still, the US government could not abandon seven heroes. Within hours of the explosion, the navy launched the largest salvage operation in American history. NASA officials swore they would do everything possible to retrieve the crew's bodies and bring them back to their families.

—◆—

Training director Hughes had watched the launch from a colleague's office at JSC. "Three, two, one, zero, they're going up, and then they were gone." It took Hughes a minute to remember the crew list. "We were launching so often and there were so many crews in training . . . and then I think, 'Oh, Scobee. Shit.'" Everybody liked Scobee. "Then I said, 'Judy.'" Hughes had a meeting with Judy Resnik scheduled after the flight. "A brilliant person," he called her, but not the most patient person. She'd said she wanted to discuss the flight-training program, which she thought was too time-consuming, a little too dumbed-down. Next, Hughes thought of "Christa McAuliffe, because that's all people were talking about" in the run-up to the mission. And poor Greg Jarvis, "the other guy." He thought of Smith, the rookie pilot, and McNair, the sax-playing laser specialist. Then he remembered that Onizuka, who had befriended Hughes's nine-year-old son, was on *Challenger*. "Onizuka. Oh, shit."

NASA engineer Doc Pepper was watching the launch at JSC in Houston and was "quite excited, as always," when the explosion came. "For a few seconds we all just stared and did not quite understand what we were looking at. What threw me is that the incoming data kept coming in and the astronauts' comm talk continued. All that soon stopped and we began to realize something was definitely wrong. I was as stunned as everyone else. Looking back, I'm not sure it was the astronauts actually talking on the comm, or Mission Control, but for years all I could think about was that at that moment, I was listening to ghosts."

The agency's harried PAOs filed a message from Vermont senator Patrick Leahy that morning. Senator Leahy wanted some answers about the "shuttle launch delays" that had been embarrassing the agency and its congressional supporters. His request was logged into the public affairs office at 11:54 a.m. on January 28, fifteen minutes after *Challenger* burst into pieces. A note stapled to it later read, "*Leahy—Shuttle Launch Delay . . . forget it.*"

In Florida, Johnny Corlew, the closeout-crew chief who had offered Christa an apple in the White Room, still had the apple. "I went home and told my wife, 'I don't want to talk to nobody,'" Corlew recalled. "I went back to work the next day and told my boss, 'Put me on annual leave.' I went to Lake Okeechobee and fished for three weeks. I couldn't do it anymore after *Challenger*."

—◆—

On the day of the explosion, Concord High School senior Matt Mead received a letter from Florida. It was the letter of recommendation Christa had written for him on the flight from Houston to Cape Canaveral. Mead remembered "Mrs. McAuliffe telling us how important it is to go out and experience new things. Even if it's scary, even if it means moving away from the people you love and care about, she said you have to do it."

Principal Foley canceled classes for a day and a half. The football team helped him keep reporters out of the school. "We're going to talk among ourselves. We're going to lean on one another," Foley said. He brought in a team of grief counselors to monitor students for signs of depression or post-traumatic shock. In the end the counselors outnumbered the kids who turned to them for help. "We felt awful for Mrs. McAuliffe," one member of the Concord High class of '86 recalls, "but what could we do? What did we have to do with it? I don't want to sound cold, but we were teenagers thinking about prom and jobs and college." Many of his classmates ignored the professional counselors and turned to a peer-counseling group called Teens Need Teens instead. "We figured adults only screwed things up."

Outside the school's redbrick walls, Concord was reeling. "The city was numb," says Mike Pride, editor of the *Monitor* at the time. Pride had been at the local courthouse "when somebody ran in with the news. '*Challenger* exploded!'" He ran a block and a half to his newsroom, where his young staff looked as numb as he felt. "We'd spent six months covering this wonderful story, and now the story was *grief*." Pride told his staffers to report the news while respecting their neighbors' feelings, but the out-of-town press had no such compunctions. "We saw a lot of bad behavior. There was a church service where national photographers took over the first row and shot backward" at Christa's grieving friends and neighbors. "When Scott McAuliffe's third-grade class came home from Florida, some of the kids were crying as they got off the bus from the airport—with flashbulbs going off in their faces."

Out-of-town reporters bought plane tickets that got them into the arrival area at Boston's Logan Airport, where they waited for the Corrigans' flight from Orlando. Christa's weary, shell-shocked family dodged the newshounds by following a state trooper who met them on the tarmac and drove them home to Framingham in an unmarked car. "We had been home only a few hours when the doorbell rang," Grace recalled. "A reporter from the *National Enquirer* was standing there when our son opened the door. 'We'll give you $100,000 for an exclusive interview,' he said."

Christa's brother shut the door in the man's face.

——◆——

June Scobee flew home to Houston with her children. For her, a thousand miles wasn't enough distance from the Cape. "It was like we never left. We couldn't turn on a TV without seeing the accident over and over." There was something peculiarly new to that. As CBS anchor Dan Rather said later, "This event set a precedent for coverage in which the video was so spectacular, so tragedy-laden, that we repeated it over and over again." A younger generation would experience another televised trauma, the September 11, 2001 attacks on the World Trade Center, in much the same way.

Astronaut Mullane and his wife, Donna, "drove to visit the widows" the day after the accident. "We first went to June Scobee's home. The street in front of her house was a mob scene. A large crowd of the curious filled the neighbors' driveways and lawns." They found the Scobees' house "filled with family, friends, and several astronauts and their wives. June was the picture of exhaustion, her face puffy and tear-stained." Astronaut

Norm Thagard was staying with June and her kids, fielding phone calls for them.

For days, the mission's mother hen played hostess to dozens of visitors. "That's what Dick would have wanted. But in my mind I wasn't a widow yet. I was still Dick Scobee's wife. I didn't accept that he wasn't coming home until the first time I went to the grocery store. He loved peanut butter, and I picked up a jar of peanut butter like I always did. Then I sat down and cried, because I didn't have him to take that peanut butter home to."

—◆◆—

Americans had their own troubles processing the disaster. One of the earliest reactions was a series of dead-astronaut jokes. Psychologists described them as gallows humor, a defense mechanism. "Anytime something stirs the larger cultural and national emotions, there will be a joke cycle," one pundit explained. "I heard shuttle disaster jokes within twenty minutes." Many of the jokes were unprintably cruel, but one that played on the agency's love of acronyms made a sort of sense: "What does NASA stand for? 'Need Another Seven Astronauts.'"

But perhaps there was meaning to be found in the *Challenger* story. Nationally syndicated humorist Art Buchwald, who had spoofed the Teacher in Space program in a prelaunch column, turned serious after the accident: "Seven brave people died that morning, but it was the death of a schoolteacher that made our children cry." Recalling how the space race had spurred the United States to "make our schools second to none," Buchwald lamented a long slide that was turning teachers into underpaid service workers. Then, in "one horrifying moment in Florida, things changed again. The parent-teacher-pupil bond that had been fraying seemed to be joined again." Buchwald noted that

Christa planned to teach again when she got back from space. She wanted to promote the cause of her fellow schoolteachers. Now, wrote Buchwald, "Thanks to Christa, each one of them can say with pride, 'I'm a teacher too.'"

Michael Collins spoke for many NASA veterans in a *Washington Post* op-ed published less than forty-eight hours after the disaster. "I have been expecting something like this for more than twenty years," wrote Collins, who had flown with Armstrong and Aldrin on Apollo 11 and advised Resnik on her efforts to become an astronaut. To him, Christa's involvement complicated matters because she hadn't spent years living with the risks of spaceflight. Collins compared the accident to the *Hindenburg* and to a 1985 Japan Airlines crash that killed 520. "One difference this time is the eye of the television camera. We watch the rocket explode, over and over again, in slow motion, backward and forward, in vivid color. Christa's parents watch their daughter's spacecraft being blown to bits. We watch them watching it. Not once but as often as we can stomach it." In the end, Collins saw the loss of the shuttle and its crew as a chance to learn a lesson. According to him, "the only way to be 100 percent sure of avoiding air accidents is to stay on the ground." Asking if there was a moral to the *Challenger* story, he wrote, "I think I do know what my friend, astronaut Judy Resnik, would say: fix the problem, and then let's get on with it."

That Friday, Ronald and Nancy Reagan flew to Houston for a memorial service to honor the *Challenger* crew.

Reagan hated funerals. "He doesn't go to the funerals of his friends," recalled speechwriter Noonan. But this was different. Three days after the explosion, the president wanted to follow his

Tuesday-night speech to the nation by eulogizing the Challenger Seven.

Before the service, Reagan met privately with the crew's families.

"My husband would be so honored," June said when Reagan clasped her hand in both of his. Mustering a smile, she said Dick had told her that flying was worth the risk. He wouldn't have wanted his death to spell the end of the shuttle program. "He'd want it to be bigger and better and safer." Then the commander's wife added a quibble. "After his first spaceflight," June told the president, "you went on TV and mentioned them all *except* Dick Scobee!"

Two thousand family members, astronauts, congressmen, and assorted VIPs sat on folding chairs on the lawn outside the avionics building at JSC, where the agency traditionally held welcome-home ceremonies. Another eight thousand filled a standing-room area at the end of the lawn.

Nancy Reagan took a seat in the front row between June Scobee and Jane Smith. She stood up to embrace Grace Corrigan, giving Grace a kiss on the cheek. "You've lost a beautiful daughter," the First Lady said.

President Reagan shook Ed Corrigan's hand before he went to the podium overlooking the crowd.

"We come together today to mourn the loss of seven brave Americans, to share the grief that we all feel, and, perhaps in that sharing, to find the strength to bear our sorrow and the courage to look for the seeds of hope," he began. "The best we can do is remember our seven astronauts, our Challenger Seven—remember them as they lived, bringing life and love and joy to those who knew them, and pride to a nation."

He named them. "Dick Scobee," the president said. "He

served as a fighter pilot in Vietnam, earning many medals for bravery, and later as a test pilot of advanced aircraft before joining the space program. Danger was a familiar companion to Commander Scobee."

Reagan said, "We remember Michael Smith, who earned enough medals as a combat pilot to cover his chest, including the Navy Distinguished Cross, three Air Medals, and the Vietnamese Cross of Gallantry with Silver Star in gratitude from a nation he fought to keep free." Fourteen-year-old Allison Smith wept through the president's speech.

"We remember Judith Resnik," Reagan said, "known as J.R. to her friends, always smiling, always eager to make a contribution, finding beauty in the music she played on her piano in her off-hours.

"We remember Ellison Onizuka, who as a child running barefoot through the coffee fields and macadamia groves of Hawaii dreamed of someday traveling to the moon. . . .

"We remember Ronald McNair, who said that he learned perseverance in the cotton fields of South Carolina. His dream was to live aboard the space station, performing experiments and playing his saxophone in the weightlessness of space." That wasn't quite true; McNair was going to leave NASA after *Challenger* landed. Still, Cheryl McNair was pleased to hear the president say, "Well, Ron, we will miss your saxophone, and we will build your space station."

Jarvis got relatively short shrift as usual. "We remember Gregory Jarvis," the president said. "On that ill-fated flight he was carrying with him a flag of his university in Buffalo, New York—a small token, he said, to the people who unlocked his future.

"And we remember Christa McAuliffe," Reagan went on, his voice rising, "who captured the imagination of an entire nation, inspiring us with her pluck, her relentless spirit of discovery. A teacher—not just to her students but to an entire people, instilling us all with the excitement of this journey we ride into the future."

He paraphrased Christa: "Sometimes, when we reach for the stars, we fall short, but we must pick ourselves up again and press on."

At the end of his remarks, the president mentioned Dick Scobee not once but three times. "Dick Scobee knew that every launching of a space shuttle is a technological miracle," Reagan said, "and he said, 'If something ever does go wrong, I hope that doesn't mean the end to the space shuttle program.' Today we promise Dick Scobee and his crew that their dream lives on." In closing he said, "Dick, Mike, Judy, El, Ron, Greg, and Christa, your families and your country mourn your passing. We bid you goodbye. We will never forget you."

A band from Lackland Air Force Base played "God Bless America" while NASA T-38s flew in formation overhead. They performed what is called the "missing man" formation, in which one jet peels off at the last moment, leaving an empty space.

June Scobee had barely slept during the worst three days of her life. "I burst out in tears," she says. "Nancy Reagan reached over and squeezed my hand."

After the memorial, Massachusetts senator Edward Kennedy met privately with the McAuliffes and Corrigans. Kennedy introduced them to his niece and nephew Caroline and John. Many Americans still thought of Caroline and "John-John" Kennedy as brave preschoolers at their father's 1963 funeral at Arlington National Cemetery. JFK Jr. was now twenty-five years old and six feet tall, a community organizer in New York, and twenty-eight-year-old Caroline was a law student at Columbia.

"Your daughter was an inspiration to me," Caroline Kennedy told Christa's parents.

"And your father inspired Christa," Ed Corrigan replied.

——◆——

After they flew back to New England, Steve McAuliffe asked his mother-in-law if she and Ed would stay with him and the kids in Concord for a while. When Grace asked how long they should stay, Steve managed a smile. "How about ten years?"

He gave the Corrigans the third-floor bedroom with its Jacuzzi, skylight, and the drapes Christa had hung in December. Steve bunked downstairs with the children. Over the next few days Grace cooked and cleaned, answered the phone, and shooed reporters out of the driveway. She helped handle cards, letters, and flowers that flooded in from all over the world, such an out-pouring of grief and sympathy that the local post office couldn't contain it. The Concord Police Department set aside a back room to hold all the Christa McAuliffe mail. The room soon overflowed. Grace kept every letter, vowing to answer them all. One schoolboy's note made her laugh while practically breaking her heart. "I want to be a teacher," the boy wrote, "but it's too dangerous!"

Grace sent almost all the flowers to local hospitals, nursing homes, and churches. She kept one arrangement—a large red, white, and blue wreath from the Reagans that took pride of place on the altar at a funeral Mass held on the third of February at St. Peter's in Concord, where Christa had taught catechism classes. Cardinal Bernard Law, the archbishop of Boston, celebrated the Mass along with Father James Leary, a cousin of Christa's who had become a priest. There was no press allowed. Steve had written a speech but didn't feel up to delivering it, so his law partner Michael Callahan read it for him. Steve's eulogy began with the moment he first saw Christa at Marian High School: "She came into class, and I knew I had to meet her."

—◆—

Monitor editor Mike Pride understood the McAuliffes' desire to keep their grief to themselves, "but it robbed the city of closure. Concord was reeling." Pride's readers wanted to know how Christa's family was coping. "We wanted to be discreet and respectful, but it was our job to cover the story. So we staked out funeral homes. We staked out the cemetery." There was a plot for Christa at Calvary Cemetery, but no marker or grave. There was no body to bury.

The week after the accident, NASA released a statement from the crew's next of kin: "The 51L families want to thank the people of our country and all the countries of the world for their thoughts, their feelings and words of encouragement." As for remembrance, the families urged Americans to look to the stars. "Space flight serves as an outlet for our human need to learn and expand. What's out there will make our lives better on Earth and help satisfy mankind's natural curiosity to explore and push the borders of the 'known universe.' So that their lives were not lost in vain we *must* rededicate ourselves to the exploration of space and *keep the dream alive*."

Millions honored the Challenger Seven in one way or another. A survey reported that "over ten million American adults attended a local memorial service in honor of the astronauts." One memorial in Alabama would stand out in retrospect. In February, Marshall Space Center director William Lucas spoke to a crowd of two thousand that included Governor George Wallace and Huntsville mayor Joe Davis. A graying Tennessean, Lucas ruled the sprawling campus that hosted NASA's rocketry division with what was often described as an iron hand. Everyone who worked for him knew that Lucas, an efficiency expert who had worked with Wernher von Braun in the "Hunsville" days, did not tolerate delays. If an underling said he or she wasn't ready to sign off on a project, his answer was, "Your job is to be ready."

Now Lucas addressed a disaster that appeared to be due to one of his rockets. "Governor Wallace, Mayor Davis, distinguished guests, ladies and gentlemen," he began. "Throughout history, brave men and women have often paid a fearsome price as they pushed back frontiers or sailed on uncharted seas. This past week seven men and women, among the bravest and best of today's pioneers, were taken from us in one shocking moment." In his view, apparently, the shuttle had been downed by a bolt from the blue, not a faulty booster built and maintained by a rocketry division in which delays were not acceptable.

"If the crew of *Challenger* could speak to us today, they would undoubtedly say, 'Keep the faith. Press on to finish the task,'" Lucas announced, calling on Marshall Space Center workers and all Americans to honor the crew's "supreme sacrifice" by supporting the space program. "And with the help of Almighty God, we will not allow their sacrifice to have been in vain."

Lucas's hopeful tone matched that of many other eulogies in the days after the disaster. But soon a darker view emerged.

15

THE BIGGEST SALVAGE EFFORT IN NAVY HISTORY COVERED SIX thousand square miles from Port Canaveral north to Daytona Beach, from the shore to an imaginary line eighteen nautical miles offshore, where the water was two miles deep. More than two thousand servicemen and women took part in an operation run by Dale Uhler, the navy's deputy supervisor of salvage, whose fleet included more than two dozen ships, a nuclear-powered submarine, four search planes, and seven helicopters. Uhler was particularly keen on *Sprint*, a robotic submersible that prowled the sea floor, bristling with eyes. He debriefed the press on its "triplex camera head system, low-light black-and-white camera, color video camera, and 250-frame-capacity still camera." A NASA press release noted Uhler's estimate of 1,600 pounds of *Challenger* debris recovered so far, "most of it found floating on the surface of the Atlantic Ocean. The largest piece was approximately thirty feet by five feet. The recovery team reported it was 'aluminum-like with wires attached.'" Sailors also recovered "two cone-shaped objects with parachutes attached." Those were the noses of the rocket boosters. The boosters had parachutes, even if the crew did not.

Bits and pieces of the orbiter and the rest of the stack surfaced for weeks. Some sections of fuselage floated offshore. Others washed up on beaches north of Cape Canaveral. Salvage ships eventually recovered more than fifteen tons of debris that was laid out on the floor of Hangar L, near the Vehicle Assembly Building at KSC. NASA workers measured, photographed, and catalogued each charred chunk of fuselage, each engine component, each piece of a wheel assembly that had been meant to touch down at KSC's shuttle runway on February third. They positioned each piece on the floor of the cavernous hangar in a grid that corresponded to the National Transportation Safety Board's guidelines for airplane crashes. Still there was no trace of the crew cabin. Weeks passed with no sign of the flight deck, the middeck, or the seven people inside.

Investigators studying videotape of the orbiter's breakup picked out one piece of "ejecta" they considered a likely candidate. Object D, they called it. "The physical extent of the object was estimated from the TV recording to be about five meters," they reported. That was the size of the crew cabin. The video showed Object D falling twelve miles before it crashed into the sea at 207 miles per hour. Comparing its trajectory to radar readings, searchers zeroed in on several square miles of choppy ocean twenty miles off Cape Canaveral. Still they found nothing. Bob Crippen, who had piloted the first shuttle and commanded *Challenger* in 1984 with Scobee in the pilot's seat, grew so frustrated by the lack of results that he used his personal credit card to hire a scallop boat that trawled the sea floor off the Cape, searching for Scobee and his crew.

The *New York Times* revealed in February that NASA was aware that it might be playing chicken with shuttle crews' lives. One of the agency's budget analysts, Richard Cook, had spent months warning his superiors about the rocket boosters. Cook had huddled with engineers who shared his concerns. In 1985, he fired off a memo describing the booster as "a potentially major problem affecting flight safety." His superiors agreed to study the problem, but not to let it delay any missions. They had sixteen shuttle flights on the docket for 1986.

There were other hints that shuttle launches were accidents waiting to happen. Allan McDonald, director of the space-shuttle rocket program at Morton Thiokol, had received a memo from one of his engineers. Typed on Morton Thiokol letterhead and dated October 1, 1985, Bob Ebeling's memo started with an all-caps plea: "*HELP!*" Three and a half months before *Challenger*'s launch, it described the engineers' efforts to get company support for studying problems with the O-rings. Its last line read: *This is a red flag*.

McDonald thought Ebeling was onto something. The rockets their company sold NASA had a flaw that might make a shuttle blow up on the launchpad.

In March, six weeks after the accident, members of the presidential commission investigating the accident received an anonymous letter in a plain brown envelope. The writer, a middle manager at NASA's Marshall Space Flight Center in Alabama, called himself "Apocalypse." He blamed the rocket program's go-at-all-costs ethos on "one man, William R. Lucas," the Huntsville czar who had eulogized the Challenger Seven a month before, vowing that their "supreme sacrifice" should spur a new commitment to the shuttle program. Lucas's anonymous critic described his management style as feudalistic. "In his ten-year tenure as

Center Director, he established a personal empire. The only criterion for advancement is total loyalty to this man." Loyalty to country, NASA, or the space program "means nothing." As a consequence, the whistleblower wrote, the flight-readiness review that preceded every launch had less to do with flight readiness than with pleasing Lucas. "It was not determining flight readiness." Instead, he charged, the climate of fear around Lucas ensured that no one at a preflight review dared to say his or her team was not ready to launch. "To do so would invite the question, 'If you are not ready, then why are you not doing your job?'" Describing the atmosphere at Huntsville as "Orwellian," he claimed that "Lucas made it known that, under no circumstances, is the Marshall Center to be the cause for delaying a launch."

The letter was signed, "God help us all, Apocalypse."

Along with the *Times* story and other press accounts that began to appear, the so-called Apocalypse Letter troubled Christa's family. According to *Monitor* reporter Ray Duckler, "As the story came out, the family got more and more upset. They never publicly pointed a finger, but their refusal to label the incident an 'accident' spoke volumes about their anger." Christa's younger brother Kit told a TV reporter that NASA had "used" his sister to score points with the public.

Grace wasn't sure Kit should say such a thing, but her husband defended their son. Ed said, "We all do what we feel is the right thing." Before long he left Concord, with Grace staying behind to help Steve and the kids. Ed drove home to man the house in Framingham, where the portrait of Christa in her sky-blue NASA flight suit hung in the living room. He began writing notes that he kept to himself. One began, "*I have been angry since January 28, 1986, the day Christa was killed.* . . ." Another listed the names of high-ranking NASA officials.

With Grace's help, Steve McAuliffe did his best to make his

drafty three-story house a home again. After months of hearing himself described as "Mr. Mom" or "America's favorite house-husband," Steve got used to hearing the strange-sounding term "widower." He tried to be honest with Scott and Caroline, to let them know their mother loved them even if she couldn't come home. He shielded them from news crews that camped at the foot of the driveway and the endless TV replays of the explosion.

"I think Steve feels burned. By NASA. By the press. He's furious," a friend said. "For a while he felt guilty because he encouraged her to do it, as though he somehow should have known better. But who could know?"

Others were furious, too. In Utah, near the factory where Morton Thiokol built rockets for the shuttle program, the firm's executives and engineers drove to work under a railroad overpass leading to Thiokol's rocket-booster factory. As news of the boosters' role in the *Challenger* explosion spread, a spray-painted message appeared on the overpass:

MORTON THIOKOL MURDERERS

—◆—

The rocket makers of Utah had been in business since 1926, when Missouri chemist Joseph Patrick and his partner, Nathan Mnookin, set out to concoct an antifreeze for Henry Ford's mass-produced automobiles. They failed, but one of their experiments produced a foul-smelling synthetic rubber they called *thiokol*, a combination of Greek words for sulfur and glue. Twenty years later, an engineer at the Jet Propulsion Laboratory discovered that thiokol burned more efficiently than other early rocket fuels like gunpowder and asphalt. Thiokol bid for NASA contracts in hopes of powering the space program, but lost out when the agency opted for liquid fuel in the rockets that boosted Mercury, Gemini, and Apollo astronauts into space. Liquid fuel was safer.

Other government contracts filled the gap. Thiokol built the rockets that fueled the United States' nuclear arsenal—Minuteman and Peacekeeper ICBMs in underground silos as well as submarine-launched Poseidon and Trident missiles—and in 1973 won the right to produce the space shuttles' rocket boosters. Nine years later, it merged with Morton Salt and became Morton Thiokol, a conglomerate that billed its billion-dollar businesses as "Salt, Specialty Chemicals, and Aerospace." By 1985, the firm's aerospace division accounted for almost half its annual profits of $198 million, on sales of almost $2 billion. Morton Thiokol's 1985 annual report showed a shuttle launch over the caption, *"Morton Thiokol solid propellant rocket propulsion motors have performed flawlessly for nineteen Space Shuttle flights."* That wasn't quite true, but by January 1986, the company's rockets had boosted twenty-four shuttle crews into orbit without harming them. When the men who ran NASA needed rocket power at a fair price, they knew where to turn.

Morton Thiokol's sprawling complex outside Brigham City, Utah, which the *New York Times* described as "the world's largest rocket factory," covered thirty square miles of sagebrush flats an hour's drive north of Salt Lake City. It was here that Thiokol chemists and engineers mixed ammonium salts, aluminum powder, iron powder, and a synthetic binder in six-hundred-gallon bowls, brewing a mixture so explosive that each rocket booster could produce more power than half a dozen Hoover Dams. (One NASA gearhead preferred to say that each rocket booster packed "15,400,000 horsepower, or roughly as much as 64,000 Corvettes.") But mixing and curing such a combustible brew could backfire on its makers. In 1984, a booster segment caught fire in mid-manufacture. "All the workers were able to evacuate before the ignited segment caused an explosion, completely destroying the casting building, all the steel beams, and a two-hundred-ton overhead crane assembly," booster-program director McDonald recalled. "Pieces of these huge steel structures flew over the heads of the workers running from the building to dirt bunkers located nearly fifty yards away. It was an absolute miracle that no one was killed." A year later, a lightning strike ignited seven thousand pounds of propellant. Seven thousand pounds constituted less than 1 percent of a rocket booster's contents, but the fire destroyed another building. "Once again we were lucky that no one was killed."

Each rocket booster was made of five coffee-can-shaped segments filled with propellant. Before each shuttle mission, Thiokol's crane operators loaded booster segments onto railroad cars that rode America's crumbling rail lines 2,500 miles from Utah to Cape Canaveral. The booster segments bore a warning: DO NOT HUMP. "Hump" is a railroad term for letting freight cars roll downhill under their own power—a bad idea for a freight car carrying a hundred tons of rocket fuel. Once the segments reached the Vehicle Assembly Building at KSC, NASA workers stacked them into cylinders 150 feet high, assembling the boosters that provided the lion's share of each space shuttle's thrust. The segments were connected by rubber rings—the O-rings—that had to provide a seamless seal to contain the explosive fuel inside.

McDonald and several Thiokol engineers had growing doubts about the O-rings that held their boosters together. They had seen burn marks on O-rings recovered from previous missions. NASA knew about the burn marks, which suggested that several shuttle flights may have barely escaped an explosion on the launchpad. During the winter of 1985–86, the company appointed a task force to study the O-ring problem.

—◆—

On March 7, five and a half weeks after the disaster, navy divers from the salvage ship USS *Preserver* found the crew cabin twenty-one miles northeast of the Cape. They found it misshapen and torn—"like crushed aluminum foil," one diver said—and covered with barnacles in eighty-seven feet of murky water.

"It was just a pile of rubble," said navy diver Thomas Stock, one of the first to peer into the cabin's remains. The flight deck and middeck had been smashed on impact with the sea, steel seats torn loose, control panels crushed, lockers burst open. "We're talking debris and not a crew compartment," a navy spokesperson said, describing "remains, not bodies."

The cabin had lingered in the sea for thirty-nine days. One diver retrieved a necklace and a few long strands of dark hair. According to a longtime NASA insider, the impact with the ocean "had ripped out part of the cabin, and one of the crew was found outside the cabin. It was Judy. She went straight to the bottom, still in her seat."

As NBC's Jay Barbree reported, "The cabin wreckage was so twisted and tangled, sharp edges jutting everywhere like knife points, that the divers demanded the wreckage itself be hauled to the surface and the operation continued on deck." In the end only four of the crew's seven PEAPs, their emergency air packs, were recovered. Three of the four had been activated, proving that at least three crew members had spent their last minutes fighting for their lives.

During the last stage of the salvage operation, as a crane on the *Preserver* lifted the cabin from the sea floor, "a blue astronaut jumpsuit bobbed to the surface," Barbree reported. The ghostly blue-suited body "turned slowly, and then disappeared. Divers and sailors stood stunned as they realized what had happened." They had found—and then lost—one of the Challenger Seven. "Immediately the divers went deep again, beginning a frantic search for the last astronaut of *Challenger*." In the following days, pathologists identified remains of crew members Scobee, Smith, Resnik, Onizuka, McNair, and McAuliffe—all but Greg Jarvis, whose father complained that NASA had informed the other families that their loved ones had been found. "I'd just like to know about it, too," Bruce Jarvis told ABC News. No one told him that the divers had seen his son's body only to let it slip away. It would be another five weeks before they re-recovered Jarvis and NASA announced that all seven members of the crew had been found.

Reporters and camera crews worked overtime to cover the recovery effort. All three TV networks hired boats to follow the salvage ships. ABC hired a helicopter to watch from above. NBC put a camera crew on the jetty at the mouth of Port Canaveral to zoom in on salvage ships as they returned. One day the TV cameras filmed members of the *Preserver*'s crew on deck in their dress-blue uniforms, saluting coffins draped in American flags. NASA was not pleased. *Preserver* captain Robert Honey was told to discontinue ceremonies that "drew too much media attention."

Astronaut John Young, commander of the first shuttle mission, had watched *Challenger* go up in flames from his seat in the cockpit of a NASA pursuit plane. Young was a straight shooter who didn't mind airing his doubts about the shuttle program, at least inside the Astronaut Office. "If we do not consider Flight Safety first all the time at all levels of NASA, this machinery and this program will not make it," he wrote in an eyes-only memo in March. "There is only one driving reason that such a potentially dangerous system would ever be allowed to fly—launch-schedule pressure." After his warning was leaked to the press, Young was relieved of his duties as head of the Astronaut Office and reassigned to the do-nothing post of technical advisor. Asked to explain the agency's actions, he said, "NASA is protecting NASA."

The agency asked other astronauts to take part in *Challenger*'s postmortem. One of the most difficult tasks fell to Rhea Seddon, an MD who had worked in emergency rooms and performed autopsies in her native Tennessee. Seddon, who had become the fifth female astronaut after Ride, Resnik, Kathy Sullivan, and Anna Fisher when she flew on *Discovery* in 1985, reported to a Hangar L at KSC to examine her colleagues' remains. Her assignment was kept from the press.

Dr. Seddon thought she was up to the task of examining what was left of the crew cabin. Its fishy odor, a reminder that it had been underwater for almost six weeks, unsettled her at first. "After the initial shock of the odor," she wrote later, "I managed to assume an air of detachment."

She saw the familiar detritus of spaceflight: food packs, serving trays, spoons and forks, checklists, pencils. She saw a deflated soccer ball, a souvenir Onizuka had promised to bring home to the schoolchildren he coached in Hawaii. Then she lifted a corner of tarp that covered a box of recovered debris. "I was taken aback by a most surprising sight. Smiling up at me was a picture of Colonel Sanders, of fried-chicken fame." Colonel Sanders's familiar face marked what was left of an experiment sponsored by Kentucky Fried Chicken that would test how fertilized chicken eggs developed in the microgravity of Earth orbit.

There were also "panels of switches and controls from the flight deck." Seddon saw that some of the switches were "out of configuration—that is, not where they should have been during the launch. Had they been knocked awry during the explosion and crash?" she asked herself. "Was there any logic to where they were now?" At the time she doubted that her friends could have been aware of their plight after the explosion. But there were indications that they had been. For one thing, some of the switches were guarded by spring-loaded lever locks. It took a careful two-step process to unlock and turn them. For another, Seddon noted, "If the cabin walls had been split open during the explosion, there might be signs of hydrazine fuel from the nearby reaction-control jet tanks in the human tissues." But there was no hydrazine in what was left of the Challenger Seven. That made it more likely that Christa and the others had been alive, aware, and even active during the two minutes and forty-five seconds between the explosion of the external tank and the moment the crew cabin crashed into the ocean at 207 miles per hour.

Seddon had a harder time maintaining her professional detachment on her second trip to the hangar. "Jane Smith," the pilot's widow, "made the simple request that her husband be buried in the NASA flight suit he had been wearing. I was asked to retrieve Mike's clothing and make sure it got cleaned up to send to the funeral home, even though the casket would remain closed." Seddon retrieved Smith's sky-blue flight suit from the carefully catalogued debris of the wreckage, "grimy and grungy from the destruction of the crash and its long stay in the sea." She carried what was left of the suit to a part of the hangar that was serving as a morgue. Using detergent and a scrub brush, she washed the fabric until her knuckles were raw. "Soon tears began pouring down my face, splashing into the wash water."

Smith and Dick Scobee were buried in separate plots among the rows of white gravestones at Arlington National Cemetery. Smith was buried on the third of May, 1986, and Scobee on May 19, which would have been his forty-seventh birthday. Ron McNair was buried in his native Lake City, South Carolina. Ellison Onizuka was buried in the National Memorial Cemetery of the Pacific, a green volcanic crater on Oahu. Resnik's and Jarvis's families took their remains home to Ohio and California, respectively. At a memorial service for Resnik in Akron, female astronauts including Sally Ride waved to a flyover by four NASA jets, using the American Sign Language signs for "I love you." Greg Jarvis was cremated, his ashes cast into the Pacific near a spot he used to surf.

Christa McAuliffe was buried on a hillside at Calvary Cemetery in Concord on May 1, 1986. Her grave was unmarked. Her husband didn't want tourists to know where it was. Steve McAuliffe and his children joined a few friends and family members for a graveside service presided over by Father James Leary, Christa's cousin, who had married Christa and Steve in 1970.

——◆——

Not everything recovered from the crew cabin could be identified. A box containing what the *New York Times* described as "a small residue of remains" was buried at Arlington National under a stone that holds a bronze plaque showing the Challenger Seven. "We buried part of him with the others," June Scobee says of her husband. Commander Scobee's astronaut pin went into the box "because he would have liked that."

In July, Joseph Kerwin, the former shuttle astronaut serving as medical director at JSC, announced the results of the agency's postmortem: it was inconclusive. Due to the force of the crew cabin's crash into the sea, Kerwin's team "could not determine whether in-flight lack of oxygen occurred, nor could they determine the cause of death."

Those issues would occupy a presidential commission for four months. Chaired by William Rogers, who had served as President Eisenhower's attorney general and Nixon's secretary of state, President Reagan's blue-ribbon panel featured Chuck Yeager, the flyer who broke the sound barrier in a rocket-powered jet in 1947, as well as Neil Armstrong, Sally Ride, and a shaggy haired, bongo-playing, Nobel Prize–winning physicist who suspected that NASA was full of baloney.

16

RICHARD FEYNMAN'S PHONE RANG. THE CALLER WAS WILLIAM
Graham, a former student of his at Caltech, now acting director
of NASA. Feynman didn't remember Graham and didn't like the
sound of what he was calling to offer: a seat on the Presidential
Commission on the Space Shuttle *Challenger* Accident. Feynman
said, "You're ruining my life!"

At sixty-seven, the Nobel Prize–winning physicist was perhaps
the most famous scientist in the world. During World War II, he
had worked on the Manhattan Project that built the atom bomb.
During the late 1940s and early 1950s, he helped crack the sub-
atomic code of quantum electrodynamics, inventing "Feynman
diagrams" to show how light and matter interact. By the winter
of 1985–86, Caltech's longhaired graying eminence was happy and
comfortable in Pasadena, though he was still fighting a rare cancer

that had almost killed him eight years before, when surgeons removed a tumor larger than a grapefruit from his stomach. Feynman never saw any point in wondering if his work on the A-bomb had caused his cancer. His theoretical work suggested that time's forward motion may be little more than an illusion, a shortcut humans use to negotiate one of the universe's four dimensions, but in human affairs he never looked back.

After Graham's call he asked his wife, Gweneth, "How am I gonna get out of this?"

She urged him to join the commission. "If you don't, there will be twelve people all going around from place to place." If he joined, there would be eleven people following an itinerary like normal bureaucrats "while the twelfth one runs around all over the place, checking all kinds of unusual things. There isn't anyone who can do that like you can."

As Feynman recalled, "Being very immodest, I believed her."

He went to Washington, where Graham introduced him to Neil Armstrong—"the moon man," Feynman called him—and "the big cheeses of NASA." He met legendary test pilot Chuck Yeager, who was as uneasy in the halls of government as he was. "I had to think about whether or not to participate," Yeager admitted later. "I knew that NASA was screwing up." Feynman met their fellow commissioners: astronaut Sally Ride; diplomat David Acheson, the son of former secretary of state Dean Acheson; scientists Arthur Walker and Albert Wheelon; air force officials Eugene Covert, Alton Keel, and Donald Kutyna; *Aviation Week* editor Robert Hotz; and chairman William Rogers, who opened the hearings of what the media dubbed the Rogers Commission on February 6, 1986, nine days after the accident and more than a month before the crew cabin was found. Rogers, seventy-two, was a patrician New Yorker in a charcoal suit and a red-white-and-blue-striped tie. He had a high forehead and a level gaze that gave nothing away. Rogers also had a mandate from President Reagan.

"Whatever you do," Reagan had told him, "don't embarrass NASA."

The chairman had no plan to do so. "We are not going to conduct this investigation in a manner which would be unfairly critical of NASA," he announced at the commission's first session, "because we think—I certainly think—NASA has done an excellent job, and I think the American people do."

The first witness to appear before the commission was Graham, the agency's acting director. He raised his right hand and swore to tell the truth. A lean forty-eight-year-old with wire-rim glasses and a wispy brown mustache, Graham had been a nuclear-weapons specialist at the Rand Corporation before joining NASA. He began by addressing the commissioners. "NASA welcomes your role in reviewing and considering the facts and circumstances surrounding the accident of the space shuttle *Challenger*," he said. "You can be certain that NASA will provide you with its complete and total cooperation." That would turn out to be false.

Rogers and several other commissioners had no knowledge of aerospace matters, so a parade of agency officials followed Graham, describing how the shuttle worked. That left the scientists on the panel sitting through explanations of physics and engineering littered with what Feynman called "the crazy acronyms that NASA uses," from SRB and ET to LOX (liquid oxygen), HPFTP (high-pressure fuel turbo pump), and HPOTP (high-pressure oxygen turbo pump). Feynman complained to his wife about "how inefficient a public inquiry is: most of the time, other people are asking questions you already know the answer to." Inefficiencies drove him to distraction. "Although it *looked* like we were doing something every day in Washington, we were, in reality, sitting around doing nothing most of the time."

Feynman spent his free hours chatting with physicists at NASA headquarters on E Street, a short walk from his Washington hotel. When Rogers heard about that, the chairman issued an order barring the gadfly Nobelist from the building. Too late— Feynman had already learned what he needed to know.

He discovered that some of the agency's managers had been "fooling themselves." Asked to estimate the risk of a catastrophic accident that would destroy a space shuttle and its crew, they put the odds at 1 in 100,000. As Feynman wrote in his memoir *"What Do You Care What Other People Think?"*, that number meant that they "could launch a shuttle each day for 300 years expecting to lose only one." Engineers put the risk closer to 1 in 200, leading him to wonder, "What is the cause of management's fantastic faith in the machinery?" He was willing to bet it had to do with a logical fallacy: "NASA had developed a peculiar kind of attitude: if one of the seals leaks a little and the flight is successful, the problem isn't so serious. Try playing Russian roulette that way: you pull the trigger and the gun doesn't go off, so it must be safe to pull the trigger again."

He asked seemingly simple questions: What were the boosters' O-rings made of? Did NASA have a quality-control department? Did someone have final say on whether to launch or not to launch, or was responsibility diffused to the point that nobody could be blamed for anything in particular? But when he pressed commission witnesses for details, the chairman cut him off. One afternoon, "Mr. Rogers brought the meeting to a close while I was in midstream! He repeated his worry that we'll never really figure out what happened to the shuttle."

The commission's work gained urgency on Sunday, February 9, when the *New York Times* reported that NASA had been warned about problems with the O-rings—not only recently but for years. Later that day, NASA chief Graham treated Feynman to a movie at the Smithsonian's National Air and Space Museum. They attended a VIP showing of *The Dream Is Alive*, an IMAX film on the shuttle program. Featuring footage saved from the 1984 mission when Resnik's hair got caught in the camera, the movie "was so dramatic that I almost began to cry," Feynman remembered. As for *Challenger*, "I could see that the accident was a terrible blow. To think that so many people were working so hard to make it go—and then it busts—made me even more determined to help straighten out the problems of the shuttle as quickly as possible, to get all those people back on track." With the shuttle program on hold pending the findings of the Rogers Commission, thousands of NASA employees were eager to get back to work. "After seeing this movie," he wrote, "I was very changed, from my semi-anti-NASA attitude to a very strong pro-NASA attitude."

After the film he got another surprise phone call. Air force general Kutyna, another commission member who had become a friend, invited Feynman to his house for dinner that evening.

The general had an agenda. Earlier in the week, their fellow commissioner Sally Ride had slipped Kutyna a sheet of paper: a NASA document the agency was keeping from the press, the public, and the presidential commission. It held two columns of numbers, one showing the air temperature at previous shuttle launches, the other showing the resilience of rocket boosters' O-rings at various temperatures. The correlation was clear: the boosters' rubber O-rings didn't work as well at low temperatures. Ride, a NASA employee, was risking her job by leaking an internal document to Kutyna. He recognized its importance but couldn't reveal that it came from an astronaut. So he asked Feynman over for dinner.

After a pleasant meal the general gave the scientist a tour of his garage, which was littered with tools and auto parts. Kutyna, a car buff, had been working under the hood of a sporty Opel GT. Feynman saw the carburetor laid out on a workbench. There are several accounts of their conversation that night; in all of them, Kutyna says something like, "Professor, the rings in the engine leak when it's cold outside. Do you think cold weather might affect O-rings?"

Feynman recalled it as a head-slapping moment. "Oh!" he said. "It makes them stiff. Yes, of course!"

The next morning—Monday, February 10—the two of them stopped by Graham's office at NASA headquarters. According to Feynman, they "asked if he had any information on the effects of temperature on the O-rings." Graham said no, but promised "he would get it to us as soon as possible."

That day's hearing was closed to the press. Rogers opened by denouncing the press for revealing that NASA had ignored warnings about the O-rings. "I think it goes without saying that the article in the *New York Times* and other articles have created an unpleasant, unfortunate situation," the chairman said, adding, "There is no point in dwelling on the past." Still Rogers couldn't avoid addressing the thrust of the *Times* story: that every launch dating back to the shuttle program's first year had been an accident waiting to happen. With the press barred from that day's closed session, he invited NASA and Morton Thiokol officials to explain why.

Lawrence Mulloy, director of the agency's rocket-booster program, swore that each step of the countdown to *Challenger*'s launch followed established procedures. Mulloy, a twenty-five-year veteran of the space program, reported to Lucas—the Huntsville czar who would not take "not ready" for an answer. When Ride pressed him, asking Mulloy if he or the executives and engineers who worked for him had any concerns about the boosters' O-rings, he said, "I don't recall any."

Allan McDonald, director of the rocket-booster program at Morton Thiokol, raised his hand. In the chain of command that ran from NASA's top administrators through second-level chiefs like Lucas and third-tier executives like Mulloy, McDonald was at the level just below Mulloy. Now, he stood up. "Mister Chairman," McDonald said, "we recommended not to launch."

That got everyone's attention. As Feynman recalled, "Mr. Rogers decided that we should look carefully into Mr. McDonald's story, and get more details before we made it public. But to keep the public informed, we would have an open meeting the following day, Tuesday."

On Tuesday, Feynman woke early and hailed a cab to drive him around until he spotted a hardware store. It wasn't open yet. The Nobel Prize–winner waited in the cold "in my suit coat and tie, a costume I had assumed since I came to Washington, in order to move among the natives without being too conspicuous." When the shop opened he bought a clamp and a pair of pliers.

During Tuesday's televised hearing, Feynman pressed Mulloy about the O-rings. "If this material weren't resilient for, say, a second or two, would that be enough to be a very dangerous situation?" he asked.

"Yes, sir," Mulloy admitted.

While the hearing continued, Feynman commandeered a scale model of the space shuttle that had been passed around the room. He used his hardware-store pliers to pull a rubber strand of O-ring off the model. Then, reasoning that the temperature of the ice water that waiters and waitresses delivered to the commissioners was close to 32 degrees—a close match for the air temperature when *Challenger* launched—he dunked the chunk of rubber into his ice water. He was about to speak up when Kutyna, sitting beside him, said, "Not now." The cameras were still on Mulloy, who was droning on about the agency's preflight preparations.

Moments later, Rogers called for a recess. During the break the chairman, standing beside Neil Armstrong at a urinal in the men's room, was overheard saying, "Feynman is becoming a real pain in the ass."

When they resumed, Rogers clicked a red button on his microphone. Now he was live on national TV. "Dr. Feynman has one or two comments he would like to make," Rogers said.

Sally Ride smiled.

Feynman pressed the red button on his mic. "This is a comment for Mr. Mulloy," he said. He held up a chunk of O-ring for the TV cameras, explaining, "I took this stuff that I got out of your seal, and I put it in ice water. And I discovered that when you put some pressure on it for awhile and then undo it, it doesn't stretch back. It stays the same dimension. In other words, there is no resilience in this particular material when it is at a temperature of thirty-two degrees. I believe that has some significance for our problem."

Rogers broke in. "That is a matter we will consider in the session we will hold on the weather," he said, "and I think it is an important point, which I'm sure Mr. Mulloy acknowledges." But there was no denying the impact Feynman's demonstration had on the proceedings. His waving a chunk of chilled rubber for the cameras would be played and replayed all over the world. As Feynman's friend and fellow physicist Freeman Dyson put it, "The public saw with their own eyes how science is done, how a great scientist thinks with his hands, how nature gives a clear answer when a scientist asks her a clear question."

During three months of hearings that spring, Feynman continued his detective work between visits to a Washington hospital for cancer treatments. "I am determined to do the job of finding out what happened—let the chips fall!" he wrote to his wife. He expected the agency would try to overwhelm him "with data and details . . . so they have time to soften up dangerous witnesses, etc. But it won't work because (1) I do technical information exchange and understanding much faster than they imagine, and (2) I already smell certain rats that I will not forget, because I just love the smell of rats, for it is the spoor of exciting adventure."

—◆—

As it turned out, Feynman's adventure revolved around a teleconference Mulloy and other decision makers had held on the night before the launch. The conference began as a routine flight-readiness review, with men responsible for various facets of mission preparation weighing in by speakerphone from three conference rooms in three time zones. As Feynman observed, "It turned out that NASA's Marshall Space Center in Huntsville designed the engines, Rocketdyne built them, Lockheed wrote the instructions, and NASA's Kennedy Space Center installed them! It may be a genius system of organization, but it was a complete fuzdazzle as far as I was concerned."

The teleconference began at 8:15 p.m. Eastern time on Monday, January 27, the evening after the hatch-bolt debacle led to the shuttle program's most embarrassing scrub. Weather forecasts called for record cold at Cape Canaveral. Mulloy led a six-man contingent at the Cape. His delegation, including Morton Thiokol's Allan McDonald, sat in a trailer behind the hulking Vehicle Assembly Building. Another fifteen executives and engineers joined the call from NASA's Marshall Space Center in Huntsville, Alabama. Another fourteen checked in from a conference room at Morton Thiokol headquarters in Utah, where a microphone sat on a fifty-foot table and another mic hung from the ceiling.

Thiokol's engineers were concerned about the weather. They

came prepared with charts showing a correlation between cold weather and O-ring burn marks in the boosters recovered from previous missions. Those burn marks were warning signs. The engineers had convinced Joe Kilminster, Morton Thiokol's vice president of space booster programs, who agreed with them that *Challenger* should not launch until the temperature was at least 53 degrees, the coldest reading for any previous launch.

Mulloy considered that a hunch. The shuttle program manager subscribed to the creed posted on a wall at JSC: IN GOD WE TRUST, ALL OTHERS BRING DATA. He said the engineers' data was "inconclusive." Was he supposed to call off a half-billion-dollar shuttle launch on the basis of their worries? Where was their proof?

"My God, Thiokol, when do you want me to launch? Next April?"

George Hardy, the Marshall Space Center's deputy director, chimed in from Huntsville. Hardy said he was "appalled" at the engineers. Mulloy agreed: "The eve of a launch is a hell of a time to be generating new launch criteria."

Kilminster asked for a time-out. "Can we go offline and have a caucus?" He knew time was short. They were under pressure that mounted by the hour. Mulloy, Kilminster, and the men above them in the hierarchy, from Lucas and Graham all the way up to President Reagan, expected a Tuesday launch. Dan Rather and Peter Jennings were making fun of their instrument funnies.

With the crew and immediate family members passing the time over dinner and a few beers in their quarters at KSC, the offline caucus in Utah stretched from five minutes to twenty. Thiokol's engineers argued against a launch in record cold. Their bosses pressed them to change their minds. Until that night, each participant in each flight-readiness review was asked to affirm that as far as his department was concerned, the shuttle was safe to fly. Now, if they wanted to delay the launch until the weather at the Cape warmed up, the engineers were asked to guarantee that the shuttle was *unsafe*—that launching in the cold would definitely lead to disaster. The managers had shifted the customary presumption of danger to an assumption of safety, placing an impossible burden of proof on the engineers. As McDonald wrote later, the engineers "could not *prove* the O-rings would fail."

During the break, Jerald Mason, one of Morton Thiokol's vice presidents, turned to another, Bob Lund, who reported to him. Lund had sided with the engineers. Now his boss said, "It's time for you, Bob, to take off your engineering hat and put on your management hat."

McDonald said, "If anything happens to this launch, I wouldn't want to be the person that has to stand in front of a Board of Inquiry." But even he stood down after Lund changed his mind. So did Kilminster, who returned to the teleconference after a half-hour break by phoning Mulloy at the Cape. Sounding like a changed man, Kilminster admitted there was "concern" at Morton Thiokol about the O-rings' performance, but now he agreed that the data was inconclusive. Thiokol was go for launch.

Mulloy had the answer he wanted, but it wasn't enough. He wanted a telefax, as they were called in 1986, confirming Morton Thiokol's change of heart. This was what some NASA veterans called a CYA move, short for "cover your ass." Kilminster complied, faxing a one-page document on Thiokol letterhead recommending *LAUNCH PROCEED ON 28 JANUARY 1986*, adding a signature he would later regret. It was almost midnight when the fax reached Cape Canaveral. Mulloy showed it to a colleague, who asked, "Should we wake the old man and tell him about this?" He meant Lucas, their boss. Mulloy said, "I wouldn't want to wake him. We haven't changed anything relative to launching. If we decided to scrub, *then* we'd have to wake him."

—◆—

After news of the teleconference leaked to the press, Rogers modified his tone. During one public session, he lectured Mulloy for bullying the engineers that night. "They construed what you said to mean that you wanted them to change their minds," Rogers said, "so they were under a lot of pressure to give you the answer you wanted." Before long, the men most responsible for the launch decision lost their jobs. Before 1986 was out, Mulloy would be demoted. He and his colleagues Hardy and Mason opted for early retirement along with their boss, William Lucas. Thiokol's Kilminster was demoted and soon resigned. William Graham, NASA's acting administrator, who had invited Feynman to join the commission, resigned from the agency to become President Reagan's science advisor.

Jesse Moore, head of the shuttle program, left the agency to become president of Ball Aerospace Systems. Moore's replacement at NASA was navy admiral and two-time shuttle astronaut Richard Truly, a clean-cut Mississippian who had commanded *Challenger*'s third flight in 1983. It was Truly's job to restore the agency's good name and get space shuttles flying again.

That spring, a series of internal NASA memos tracked Truly's visits to NASA's hubs at Cape Canaveral, Huntsville, and Houston. The first memo read, Truly—*Address to employees: Returning to spaceflight—safely.* Another, logged into the file by public affairs officer Brian Welch, noted, Truly—*Visit to Houston rounds out his trips to the 3 Shuttle centers. We are going to return this great nation to space. . . . Reminded his listeners he had made 3 commitments 1) support fully the Commission 2) work w/ officials on PAO policy to see that NASA returned to bus. as usual w/national media 3) Lk fwd to future. . . . Shuttles have no problem being competitive in wld mkt. When it comes to this nation's assured access to space, the best way to go in 1986 is to get the shuttles flying. "Make clear to all our customers that we are getting back in the business. . . . Must regain credibility. Must emphasize flight safety & conservative flying but over next few yrs 'will get back into robust S.S. flying.'"*

The file's final page underlined Truly's message to NASA employees in the spring of 1986: *Bottom line . . . This won't be a namby pamby program. It's a bold business. Cannot print enough $ to make it totally risk free."* At the bottom of the page was a quote from Truly, scrawled larger than the other notes: *Only way to be perf safe is to stay on the grd & I think that's too darn safe!*

In its final report to the president, the Rogers Commission would term the *Challenger* disaster "an accident rooted in history," the history of an organization that was under pressure to meet an accelerating schedule. The commission blamed the accident on "failures in communication based on incomplete and sometimes misleading information, a conflict between engineering data and management judgments, and a NASA management structure that permitted flight safety problems to bypass key Shuttle managers."

Feynman understood the patriotic fervor behind the space program. "I worked at Los Alamos, and I experienced the tension and the pressure of everybody working together to make the atomic bomb." After the 1969 moon shot, "NASA had all these thousands of people: a big organization in Houston and a big organization in Huntsville, not to mention at Kennedy, in Florida," he recalled. "You don't want to fire people and send them out in the street, so what do you do? You have to convince Congress that there exists a project that only NASA can do." To do that, the agency "had to exaggerate how economical the shuttle would be, how often it could fly, how safe it would be and the big scientific facts that would be discovered."

A draft of the Rogers Commission's final report called for nine reforms NASA needed to implement before the shuttle program could resume. They included redesigning the rocket boosters, adding crew-escape systems, and improving NASA's management structure, internal communications, and shuttle maintenance. A tenth recommendation was different in kind. It began, *"The Commission strongly recommends that NASA continue to receive the support of the Administration and the nation. . . ."* Feynman viewed that sort of cheerleading as baloney. Some of the agency's evasions and errors were even worse. They were "crap" or "bull—," two of his stronger curses. "I could see the whitewash dripping down." After reading the commission's penultimate draft, Feynman sent Rogers a telegram: *PLEASE TAKE MY SIGNATURE OFF THE REPORT.* They negotiated; finally Feynman was allowed to add a lengthy appendix of his own to the 450-page report, while Rogers got part of his tenth recommendation, now relabeled as a "Concluding Thought":

The Commission applauds NASA's spectacular achievements of the past and anticipates impressive achievements to come. The findings and recommendations presented in this report are intended to contribute to the future NASA successes that the nation both expects and requires as the 21st century approaches.

Feynman's appendix addressed technical issues such as SRB design, NASA hardware and outmoded software (including 250,000 lines of computer code, all of which he had read), and the schedule pressures that allowed shuttles to fly in what he called "a relatively unsafe condition, with a chance of failure on the order of a percent." He lamented the space program's recent history, which "has had very unfortunate consequences, the most serious of which is to encourage ordinary citizens to fly in such a dangerous machine—as if it had attained the safety of an ordinary airliner. The astronauts, like test pilots, should know their risks, and we honor them for their courage. Who can doubt that McAuliffe was equally a person of great courage, who was closer to an awareness of the true risks than NASA management would have us believe?"

Feynman urged NASA officials to "deal in a world of reality, understanding technological weaknesses and imperfections well enough to be actively trying to eliminate them." He urged NASA to be "frank, honest, and informative" in the future. In the end, he wrote, "For a successful technology, reality must take precedence over public relations, for nature cannot be fooled."

—◆—

On June 4, two days before the Rogers Commission report was released to the press and public, the agency hosted the *Challenger* families at a hotel near NASA headquarters. Over coffee and doughnuts Chairman Rogers, Sally Ride, and Bob Crippen gave the families a preview of the commission's findings. The session was private, June Scobee recalls, "but there were reporters all over the place, climbing the balconies to get in."

She and her children still supported the space program, but this was the first time they heard details of the teleconference on the eve of the launch. Her son, Rich, a twenty-two-year-old pilot, banged his fist on a table. "Those idiots!" he said.

17

"TWO, ONE, ZERO AND *LIFTOFF*," EXCLAIMED HUGH HARRIS, the voice of Launch Control.

The date was September 29, 1988. After two and a half years, the shuttle program was back. "Americans return to space as *Discovery* clears the tower!" Harris announced over the loudspeakers at Cape Canaveral. Crowds of NASA boosters in the VIP bleachers included William Rogers, John Glenn, and congressional payload specialists Jake Garn and Bill Nelson as well as movie star John Travolta, who hoped to beat John Denver into space as a member of a shuttle crew.

Temperatures were in the eighties. Morton Thiokol had redesigned the rocket boosters for the first shuttle flight since *Challenger*, a mission the agency was calling America's "Return to

Flight." With the Space Flight Participant Program on indefinite hold, *Discovery*'s crew consisted of five professional astronauts. They wore bulky pressure suits and sixty-pound parachute packs. Their orange launch and entry suits, inevitably acronymed LESs, had been designed for what *Discovery* astronaut David Hilmers called "a *Challenger*-type accident, offering the hope that we could bail out and survive." According to Hilmers, "We knew full well that NASA's future was riding on our mission." They couldn't help thinking back to the explosion two and a half years before, especially when an alarm went off on the way up.

A false alarm, it turned out, "but you better believe it got my attention," says Hilmers. "I was seated right behind the center console on the flight deck," the seat Resnik occupied on the Teacher in Space mission.

Nobody dared to say "uh-oh" when the alarm went off. They knew that was the last thing Mike Smith said before *Challenger* broke up. Commander Rick Hauck had already chided pilot Dick Covey for saying "uh-oh" after a preflight snafu. Hauck said, "Dick, don't use those words again as long as we're flying this machine together."

The astronauts held on to their seats as the shuttle rumbled upward through turbulence. Finally, Mission Control radioed the flight deck: "*Discovery*, go at throttle up." That was another echo: pilot Covey, then serving as Mission Control's CapCom, or capsule communicator, had been the one who had radioed Scobee at the same point to say, "*Challenger*, go at throttle up."

"Roger," Hauck replied. "Go at throttle up."

After a long, tense minute, the shuttle settled into orbit. Over the next four days, the crew would deploy a satellite to replace one of the ones destroyed when *Challenger*'s cargo bay fell into the sea. Hauck, Hilmers, and the others would sleep in their zero-G bunks and wake to a blaring announcement from Robin Williams: "Goo-oo-ood morning, *Discovery*!" Astronaut Kathy Sullivan had arranged for the star of the movie *Good Morning, Vietnam* to tape a wake-up call: "Rise and shine, boys, it's time to start doin' that Shuttle Shuffle!"

On the mission's final day, the astronauts paid tribute to *Challenger*'s crew. "We couldn't let the flight pass without honoring them," Hilmers recalls, "so I sat down and tried to plan out some things we should say. We wanted people to remember their sacrifice, and why it was important for us to continue flying."

Hilmers spoke first. "We'd like to take just a few moments today to share with you some of the sights that we have been so privileged to view over the past several days. Many emotions well up in our hearts," he said as NASA TV showed the earth passing below. "Joy, for America's return to space. Gratitude for our nation's support through difficult times. Thanksgiving for the safety of our crew. Reverence for those whose sacrifice made our journey possible."

Mission specialist John Lounge added, "Gazing outside, we can understand why mankind has looked towards the heavens with awe and wonder since the dawn of human existence."

Covey spoke of technology's achievements. "As we, the crew of *Discovery*, witness this earthly splendor from America's spacecraft . . . in a fraction of a second our words reach your ears."

"Those on *Challenger* who had flown before and seen these sights," mission specialist George Nelson said, "would know the meaning of our thoughts."

That left the commander to sum up. "Today," Hauck said, "up here where the blue sky turns to black, we can say at long last to Dick, Mike, Judy, to Ron and El, and to Christa and to Greg: Dear friends, we have resumed the journey that we promised to continue for you. Dear friends, your loss has meant that we could confidently begin anew. Dear friends, your spirit and your dream are still alive in our hearts."

—◆—

During the two-and-a-half-year time-out between shuttle missions, Halley's Comet passed through Earth's orbit unobserved by Onizuka's favorite toy, the five-million-dollar satellite *Challenger* would have deployed to study the comet. The Soviet Union launched the *Mir* space station, then suffered a calamity of its own, the April 1986 explosion of the Chernobyl nuclear plant, which sent radioactive fallout as far as Alaska. In a 1987 speech in Berlin, President Reagan called on the Soviet leader: "Mister Gorbachev, tear down this wall!"

During the same period, NASA enacted most of the Rogers Commission's recommendations and quietly buried what was left of *Challenger*. After all human remains were accounted for, the agency entombed 120 tons of *Challenger* debris in a pair of unused nuclear-missile silos at Cape Canaveral. The concrete-lined Minuteman silos were eighty feet deep, with brackish water at the bottom. NASA cranes filled them with storage containers full of heat-shield tiles, chunks of fuselage, hoses and wires, parts of seats, lockers and control panels, and other remnants of the stack, along with larger pieces that wouldn't fit into the containers, and then plugged each silo with a ten-thousand-pound slab of concrete. There were no markers to show that this was where *Challenger* was laid to rest. Weeds grew up around the site, which was off-limits to the press and public. As KSC director Bob Cabana said later, "It was like they were saying, 'We want to forget about this.'"

—◆—

Many of the Challenger Seven's family members filed lawsuits against NASA and Morton Thiokol. The McAuliffe, Scobee, Onizuka, and Jarvis families chose not to sue. Relying on what they called the informal advice of one of Steve McAuliffe's law partners, those four families divided a settlement amounting to $7,735,000 in cash and annuities, with the government contributing 40 percent and Morton Thiokol paying the rest, a sum representing 3.5 percent of Thiokol's profits in 1986. (Betty Grissom, who had settled for less after the Apollo 1 fire that killed her husband, said, "It sounds like they did okay.") Thiokol also agreed to forgo ten million dollars in profits and spend more than four hundred million dollars to repair the rocket boosters it sold NASA. At the company's annual meeting that fall, Thiokol chairman Charles Locke told shareholders that the *Challenger* disaster had led to a 30 percent drop in aerospace profits in the first quarter of the fiscal year, but assured them that the future looked brighter. "Time heals all wounds," he said.

McNair's widow, Cheryl, Jarvis's father, and Resnik's mother settled a wrongful-death suit against Morton Thiokol for a sum that was not disclosed. Jane Smith sued NASA and Lawrence Mulloy, demanding a hundred thousand dollars for personal injury and fifteen million for wrongful death, citing the agency's "terrible judgments" and "shockingly sparse concern for human life." Her lawsuit noted that Mike Smith "was thrown about in the spacecraft and in the few seconds preceding his death, knew of his impending death." In a space reserved for "witnesses" to her husband's wrongful death, her attorney typed "*Several thousand people at the Kennedy Space Center.*" The pilot's widow also sued Thiokol and the United States for $1.5 billion in a case that went all the way to the Supreme Court, which upheld an appeals-court ruling that the pilot's widow could not sue because her husband had been a government employee. She would accept undisclosed settlements.

Marvin Resnik also sued over the death of his beloved *k'tanah*. Months of legal wrangling left him fuming at the way Thiokol's lawyers "wanted us to prove in court that there was pain and suffering before the crew died. Can you imagine?" Marvin was still on good terms with his former son-in-law, attorney Michael Oldak, who represented Marvin and his son, Chuck, Judy's brother. Oldak wasn't surprised by Thiokol's strategy but was taken aback by that of another lawyer he took to be lobbying against him. "Steve McAuliffe tried to keep the award to Judy's family at about one-eighth of what every other family got," he recalls. "He developed a plan where every spouse got X and every child Y." Since Resnik was single and childless, her family would receive a fraction of what other families got. Says Oldak, "There was a lot of liability to go around, but I had to fight for her father and brother." In the end, Marvin and Chuck Resnik settled independently for a sum between two million and three million dollars. Steve and his children received their share of the multimillion-dollar settlement arranged by his law partner and the payout of Christa's million-dollar life insurance policy from Lloyd's of London.

Steve sold the McAuliffes' home near Concord High and moved his family to another house nearby. He put Scott and Caroline in private school. Steve also took flying lessons. He avoided public appearances except for a speech to more than a thousand teachers at the National Education Association's national convention on the Fourth of July, 1986.

"Of all the groups in America that Christa hoped to benefit, she was most concerned about you, her fellow teachers," he began, accepting the group's annual Friend of Education award on her behalf. Some teachers in the crowd, irked by the Reagan admin-istration's cuts in education funding, wore buttons reading "Try Merit Pay in the White House" and "Budget Cuts Are a Pain in the Class." A year after becoming America's favorite househusband, Steve made it clear that he was on their side. Acknowledging that some of them had seen the Teacher in Space program as "a mere public relations ploy," he admitted their skepticism "may well have been valid." Still he saw the program as "a reasonably sin-cere effort to accomplish a national goal: to gain recognition of the basic importance of the teacher in American life. In closing, he urged the schoolteachers to "return to your states and use Chris-ta's efforts and her spirit to get involved in the political arena. To recruit and elect education candidates. To unseat those who support education with their words, but not their appropriations. And, most of all, to stay in education until we have a system that honors teachers and rewards teachers as they deserve."

That fall, he and his children returned to Calvary Cemetery for the unveiling of a large granite headstone. It was the day of the sea-son's first snowfall. Etched in white on the stone, under an image of the space shuttle, were the words:

S. CHRISTA McAULIFFE

SEPTEMBER 2, 1948–JANUARY 28, 1986

WIFE MOTHER TEACHER

PIONEER WOMAN

Steve remarried in 1992, marrying another Concord school-teacher. That fall, Republican president George H. W. Bush, recall-ing the day he named Christa America's Teacher in Space, named a lifelong Democrat to a seat on the federal bench: Judge Steven J. McAuliffe has served on the US District Court for the District of New Hampshire ever since.

He tried to give his family a normal life. "He'd play rec-league basketball with friends and go out for a few beers," says *Monitor* writer Jimenez. "He joined a bowling league, and while he wasn't great at it, he'd pretend it was really important. My sense is, not everyone recovered as well as Steve did."

Grace Corrigan celebrated her daughter's upbeat spirit in speeches at schools, libraries, and PTA meetings, and in a memoir she titled *A Journal for Christa*. Ed Corrigan's outlook was darker. He retired from his accounting job and looked after the house in Framingham, writing anguished notes his wife discovered later.

"My daughter, Christa McAuliffe, was not an astronaut—she did not die *for* NASA and the space program," he wrote. "She died *because of* NASA and its egos, marginal decisions, ignorance, and irresponsibility. NASA betrayed seven fine people. . . . I am sure that if anyone had advised her how flawed NASA management was, she would never have risked leaving her husband and children. President Reagan said that the act was not deliberate, was not criminal. But I say the sins of omission are no less sins than those of commission. I say 'they' deliberately neglected to make necessary corrections to the O-rings and are, therefore, as guilty as if they planned a deliberate criminal act." Noting that his viewpoint "differs from that of astronauts' spouses and families," he wrote, "I feel no allegiance to NASA."

"Ed wasn't happy with NASA even before the flight," says Joe Dwinell, a Framingham newspaperman who spent hours with Christa's father in those days. "He feared the worst." Grace Corrigan, on the other hand, was as sunny as their daughter. "She *glowed*," Dwinell says. Ed simmered. Poring over pictures of Christa, thinking of her, talking about her—"It's like a hole in your heart," he said. Ed fought cancer in the late 1980s, enduring chemotherapy and radiation until he told Grace he could no longer stand it. His last regret, he said, was, "I won't be with you anymore."

She said, "You'll be with Christa."

Ed Corrigan died in 1990 at the age of sixty-seven. According to one obituary, "Doctors blamed cancer, but others felt he had died of a broken heart."

—◆—

"How *could* they?" June Scobee asked herself. "How could they live with themselves?" She meant Mulloy and the other executives who had put on their manager's hats during the conference on the eve of the launch. She thought NASA had been too secretive for its own good and the good of the crews that took the risks. "If Dick knew what they knew, he wouldn't have flown. He would have said, 'Let's wait.'" The agency had invited *Discovery* commander Hauck into flight-readiness meetings before the "Return to Flight" mission—that was a step in the right direction. So were shuttle astronauts' new pressure suits and parachutes. But for the *Challenger* commander's widow all that was too little, too late. June did not return to Cape Canaveral for *Discovery*'s "Return to Flight" or the six shuttle launches that followed in 1988 and 1989.

She was touched when the International Astronomical Union named an asteroid after Dick Scobee. All six of Scobee's crewmates got an asteroid and a crater on the moon named for them, and that was only the start of the memorializing that began in 1986. Soon NASA honored Christa and Greg Jarvis with posthumous promotions from "space participant" to astronaut. California's Sunnyvale Air Force Station became Onizuka Air Force Station. Kona International Airport on Hawaii's Big Island changed its name to Ellison Onizuka Kona International; the Auburn Municipal Airport in Auburn, Washington, became Dick Scobee Field; and North Carolina's Beaufort-Morehead City

Airport became Michael J. Smith Field. It would be hard to keep track of all the schools named after the crew, a fast-growing list that included Dick Scobee Elementary in his hometown in Washington; Judith A. Resnik Elementary in her native Akron, Ohio; Gregory B. Jarvis Middle School in his hometown, Mohawk, New York; and more than forty elementary schools, middle schools, and high schools named after Christa. Cheryl McNair marveled at the transformation of parts of her husband's hometown of Lake City, South Carolina, where travelers could now turn off Ron McNair Boulevard for a visit to Ronald E. McNair Junior High or Dr. Ronald E. McNair Memorial Park, site of the Dr. Ronald E. McNair Life History Center, which occupied the old library where a librarian once called the police on nine-year-old Ron for trying to check out a science book.

In 1991, the *Challenger* families reunited for a viewing of the most imposing monument yet. The Space Mirror, rising beside a pond at the northeast corner of the Kennedy Space Center, was a slab of glistening black granite the size of an IMAX movie screen. The names of the Challenger Seven and the three Apollo 1 astronauts had been cut through the rock and backlit so that they seemed to hang in space like starlight. The nonprofit Astronauts Memorial Foundation had raised $6.2 million to build the Space Mirror, much of the money coming from sales of a Florida license plate showing *Challenger* in flight. Bruce Jarvis, Greg's father, had gotten the first license plate. Bruce had spent a day in the hospital after the accident, recovering from the shock of losing his son, but was spry enough at seventy-eight to spend hours walking through parking lots at Orlando shopping malls. Whenever he saw a car with a *Challenger* plate, he left a thank-you card under the windshield.

Massive gears built into the Space Mirror allowed it to turn and tilt like a sunflower, reflecting the sun as it passed overhead. But the gears often failed. In 1997, the pivoting monument smashed a steel beam, causing $375,000 in damage. After another malfunction brought a $700,000 repair bill, the foundation opted to spend the same money on local children. The mirror hasn't moved since. Today it faces northeast across marshland and lagoons toward the Vehicle Assembly Building and, beyond that, the launchpad where *Challenger* lifted off.

June Scobee called the Space Mirror "beautiful." She was all for any statue, school, or other memorial that reminded people of her husband and his crew, but like other members of what she called the *Challenger* family, she wondered whether that should be the end of the story. "I kept wondering if there was something Dick might have liked as much or even more." She kept thinking of a card Marvin Resnik had showed her, a sympathy card from a family friend. "*The* Challenger *is gone*," it read. "*But not the challenge.*" Some of the papers she'd found in Dick's briefcase outlined his dreams for the future. They went beyond the shuttle program to what Scobee had seen as the space program's next leaps: "new planets to explore, new worlds to build, a solar system to roam in." Quoting science fiction author Ben Bova, he wrote, "If only a tiny fraction of the human race reaches out toward space, the work they do there will totally change the lives of all the billions of humans who remain on Earth."

That gave June an idea. "I didn't want to be a weeping widow sitting at home in Houston," she recalls. She thought of MADD, the still-new organization of mothers of drunk drivers' victims, channeling their grief into good works. So June phoned Steve McAuliffe. "We'd already been talking almost every day. Steve was hurting so much." Along with Cheryl McNair, Lorna Onizuka, Jane Smith, the Resniks, and the Jarvises, she and Steve set a new goal for the *Challenger* families. As she puts it, "The world knew how our loved ones died. Our question was, Can we do something to remind the world what they lived for? Can we continue their mission?"

They decided on a teaching program. What better way to revive the Teacher in Space mission than to *teach*?

"June didn't want to build something people would walk by, nod their heads, and forget about," says a colleague who helped develop the Challenger Centers, a network of interactive campuses where schoolchildren from all backgrounds and neighborhoods learn the basics of spaceflight by playing the parts of engineers, flight controllers, and astronauts. "She was picturing a national and even international system to give kids a fun sort of STEM training. She knew Dick would have liked that. About all she had in the early days was a cardboard mockup she carried around, showing it to anybody she thought might help."

June took advantage of Vice President Bush's offer to "Call me anytime," phoning the private number on the card he gave her the night of the disaster. When he returned her call the next day, her twenty-two-year-old son answered.

"Hello," Bush said. "It's the vice president."

Rich Scobee said, "Vice president of *what*?"

After June took the phone, Bush agreed to become honorary chairman of a board of directors that included astronaut Kathy Sullivan and country singer Lee Greenwood. President Reagan pitched in by proclaiming January 28, 1987, the first anniversary of the accident, National Challenger Center Day. Soon Jim Rosebush, who had served as Nancy Reagan's chief of staff, signed on as the new project's president and helped drum up funding, including a ten-million-dollar appropriation from Congress—seed money, but not enough for the mission the families had in mind.

Bush referred June to Ross Perot. Recalling their meeting in the Texas billionaire's palatial office, she imitates Perot's adenoidal drawl. "He said, 'I don't invest in things I'm not certain are gonna pay off." He turned her down flat. "A low moment," she calls it. "I'd spent my last cash on a cab to his office, so I said, 'Will you at least give me fifty dollars to get back to the airport?' And he said, 'I don't carry cash.'"

She got along better with Walter Cronkite. The CBS anchor brightened when June came through his door with her cardboard model of what a Challenger Center might look like. Cronkite was a NASA fan with an office full of rocket models and signed photos of astronauts. He had applied for the now-cancelled Journalist in Space program. He heard her out, then asked, "How can I help?"

"We've got a dinner coming up, our first big event," she said. "How about being our emcee?"

"When is it?"

She had a quick answer to that. "When would you like it to be?"

Working around Cronkite's schedule, they scheduled a black-tie dinner for the evening of November 9, 1987.

That afternoon, June joined Jane Smith, Lorna Onizuka, Cheryl McNair, Marcia Jarvis, Grace Corrigan, and Judy Resnik's brother, Chuck, for a taping of *The Oprah Winfrey Show*. Steve McAuliffe stayed backstage. That seemed a wise choice when the interview went haywire. The more they tried to discuss their project, the more Winfrey focused on the disaster. During a commercial break, they threatened to walk off the set. A producer promised to let the "weeping widows" talk about their project, but the moment they were on-air again, Winfrey directed their attention to a TV monitor showing the explosion. "It was awful," says Rick Hutto, another White House veteran who had signed on to support the families' cause. "Oprah wanted them to narrate the deaths of their loved ones." Instead they clammed up, shaking their heads. Finally, Winfrey said, "Okay! Tell us about your Challenger Center."

When it was over, someone mentioned the Challenger Center dinner that evening. June, playing peacemaker, asked Winfrey if she'd like to attend.

Cronkite hosted the dinner, welcoming astronaut families including John and Annie Glenn. Vice President Bush gave a speech hailing June ("She's small, but she's tough") and the *Challenger* families. Winfrey arrived in time to pass out awards to Challenger Center supporters. After a dessert featuring chocolate flying saucers, Lee Greenwood sang "God Bless the USA."

"We were on our way," says June. Aerospace firms made donations to her project. The Florida legislature allotted three million dollars. The Patriots' Brian Holloway, who had framed the autograph Christa gave him, sold his Super Bowl ring for $50,000 and gave the money to the Challenger Centers. *USA Today* publisher Al Neuharth offered $250,000, which chief fundraiser Rosebush sniffed at. "That's not news," Rosebush said. "A million would be news." Neuharth agreed to quadruple his donation—if press-shy Steve McAuliffe agreed to play a round of golf with him. Steve reached for his clubs and played the only million-dollar round of his life. *USA Today* kicked in a full-page ad that spurred contributions from all over the country. After that, Rosebush drove June to a Washington, DC, post office where their foundation kept a PO box. A weary-looking clerk led them to several canvas bags full of cards and letters. Some held checks for thousands of dollars; one held a dollar bill and a note in a child's careful cursive: *From me and the tooth fairy.*

Oprah Winfrey, now a staunch supporter, hosted a fundraiser with quiz-show stars Alex Trebek, Pat Sajak, and Vanna White. When it was over, Lorna Onizuka walked past an oversized photo of the *Challenger* crew. She put a finger to her lips, touched her husband's face on the picture, and said, "Goodnight, honey."

John Denver made good on his Space Camp promise to write a song about the *Challenger* crew. Still hoping to join a shuttle crew, he performed it for the first time at a hearing of the Senate's appropriations committee chaired by former payload specialist Jake Garn. Strumming a guitar in the Senate chamber, Denver sang about spaceflight: "*I wanted to ride on that arrow of fire . . .*" The song, "Flying for Me," was dedicated to Scobee and his crew but centered on Christa. "*She gave us her spirit and all she could be. She was flying for me.*"

—◆—

The first Challenger Center opened in Houston in 1988. Dozens of schoolchildren put in three weeks of classroom prep before riding school buses to the center, where flight directors divided them into Mission Control and flight-crew teams that guided missions to the moon or Mars, or got a close-up look at Halley's Comet—not always successfully. If the student astronauts hit the wrong switches, they crashed. But here, the young flight directors gave them another life. "I watched the whole thing, clapping and crying," June says.

By 1996, there were two dozen Challenger Centers in twenty states, serving more than fifty thousand schoolchildren a year. Rich Scobee was an air force captain by then. As it happened, that January's Super Bowl between the Dallas Cowboys and Pittsburgh Steelers fell on the tenth anniversary of the *Challenger* disaster, and he was part of the pregame show. Captain Scobee joined a flight of supersonic F-16 fighter jets toward Tempe, Arizona, then peeled off at the last instant, leaving the other jets to fly over Sun Devil Stadium in the missing-man formation.

By 2003 a new, improved shuttle program had flown eighty-six missions without losing an astronaut. The shuttle seemed to be safe again.

18

THE SPACE SHUTTLE *COLUMBIA* LIFTED OFF ON THURSDAY, JAN-
uary 16, 2003. The 113th shuttle mission carried Commander Rick
Husband, pilot Willie McCool, mission specialists Michael Ander-
son, David Brown, Kalpana Chawla, and Laurel Clark, and payload
specialist Ilan Ramon, an Israeli Air Force fighter pilot, who would
be the first Israeli in space. Ramon put a microfiche copy of the
Torah in the personal preference kit in his locker on the middeck.

Columbia climbed through the usual turbulence, through
Mach One and "Go at throttle up." At T +00:01:22, with the stack
at 1,872 miles per hour, a pillow-sized chunk of insulation flaked
off the external tank and struck the orbiter's left wing, knocking
off several of the tiles that would protect the shuttle during reentry. "Foam shedding" had been happening to shuttles for years.
NASA considered it a minor risk.

Columbia spent sixteen days in orbit before heading home. The orbiter was about fifty miles up, where the sky begins to turn from black to blue, when the first alarms went off. Commander Husband switched the autopilot on and off. Pilot McCool threw flight-deck switches in an effort to regain control of the craft, but the shuttle was spinning so violently that no human could survive for more than thirty to forty seconds. The crew cabin depressurized so quickly that there wasn't time for the astronauts to shut the visors on their helmets. *Columbia* broke into pieces, disintegrating like a daytime shooting star over Texas at 7:53 a.m. on February 1, 2003.

After America's second shuttle disaster, Dick and June Scobee's daughter, Kathie, now a mother herself, wrote an open letter to the families of the *Columbia* crew on behalf of the *Challenger* children.

"We want you to know that it will be bad—very bad—for a little while, but it will get better," she wrote. "You'll torture yourself wondering if they felt pain." Seeing *Challenger* break up, she wrote, should have led to "private grief, but instead it was a very public torment. My father died a hundred times a day on televisions all across the country, and since it happened so publicly everyone felt like it happened to them, and it did. Everyone saw it, everyone hurt and everyone grieved. Everyone wanted to help, but that didn't make it easier for me. They wanted to say goodbye to American heroes, I just wanted to say goodbye to my daddy." She urged *Columbia* families to "remember the way they lived, not the way they died."

That fall, the seven names of the *Columbia* astronauts joined those of the Apollo 1 and *Challenger* crews on the Space Mirror at KSC.

—◆—

NASA's Mike Ciannilli spent several days after the accident leaning out of a helicopter over East Texas, looking for pieces of *Columbia* in the scrubland below. Ciannilli grew up in upstate New York, building scale models of shuttles, reading and rereading a popular book called *The Space Shuttle Operator's Manual.* In 1996, after waiting out a five-year hiring freeze that followed the loss of *Challenger*, Ciannilli joined NASA. He spent nine years as a test director and landing director at Cape Canaveral, preparing orbiters for launch and helping plan their smooth returns to Earth. "And then we lost *Columbia.*" He worked with NASA search crews that recovered more than eighty thousand pieces of the shuttle, each one catalogued and laid out in a gymnasium-sized room in the Vehicle Assembly Building.

NASA's official inquest recommended twenty-nine reforms to make future flights safer. After that, a question remained: What to do with the wreckage? Rather than bury it, the agency let Ciannilli walk coworkers through what he called the Columbia Room. He was an engineer, not a historian or tour guide, "but what had happened was personal—not just to me but to others who worked there. Losing *Columbia* was a gut punch." Researching the accident led him into agency archives, where he found troubling similarities between the *Challenger* and *Columbia* missions: Schedule pressures. Warning signs. He talked his way into a new project, creating a tribute to the victims of NASA's worst disasters.

"It wasn't the most-wanted job at the Cape," he recalls. "It was like showing our dirty laundry." To prepare, he read up on disasters. The *Titanic.* The *Hindenburg.* He visited the Holocaust Museum and the 9/11 Memorial in New York, asking curators how they chose their exhibits. He describes his research as "humbling." On returning to KSC he hunted down artifacts for a memorial exhibit at the Visitors' Center, where more than 1.5 million tourists per year gape at rockets, moonscapes, and the space shuttle *Atlantis.* The *Challenger* and *Columbia* families donated dozens of keepsakes he considered priceless: a TFNG T-shirt, model airplanes built by future astronauts, a *Star Trek* lunchbox, a Bible. Still he wanted more. Ciannilli wanted to display a large piece of each downed shuttle, to suggest the crafts' size and the violence it took to destroy them. He wanted to honor the forces every shuttle crew was up against. He had the empty frames of *Columbia*'s flight-deck windows, but *Challenger*'s wreckage was entombed at the bottom of those missile silos.

He found photos of several sections of *Challenger*'s fuselage. One section had been bent and burned but still showed the American flag, an image that put him in mind of the banner Francis Scott Key saw over Fort McHenry during the War of 1812, waving through rockets' glare. Thanks to the agency's detailed records, he knew which Minuteman silo that piece was in, and he convinced KSC director Bob Cabana and NASA administrator Charles Bolden to let him try to retrieve it.

Early in 2015, NASA cranes removed the concrete slab plugging one of the silos. Clamped to safety cables like a mountain climber, Ciannilli descended into the darkness, shining a flashlight around the concrete walls. He saw storage containers and rusted debris. He won't describe the smell for fear of offending the *Challenger* families, but the base of the silo was dank. He saw rusted, twisted chunks of metal. And no sign of the port-side section with the flag. He gave up for the night. This went on for more than a week. "I kept pushing for more time." After almost two weeks, he found it: a fifteen-foot section painted with a burned and scratched but unmistakable star-spangled banner. Calling for a crane, he brought the dripping chunk into the light for the first time in thirty years. When it was lying on the tarmac beside the open silo, "I went over and kissed it."

Today that charred section of *Challenger* is a centerpiece of

"Forever Remembered," the Kennedy Space Center's memorial to the fourteen astronauts who died in the line of duty during the space-shuttle era. Christa McAuliffe's Teacher in Space mission patch is on display along with Ron McNair's black belt, Ellison Onizuka's Buddhist prayer beads, a copy of Judy Resnik's journal article on frog retinas, and artifacts from the *Columbia* crew.

Sometimes Ciannilli stops by to pay his respects. "One day I heard a little girl ask her mom, 'Who were they?' And the mom said, 'They were heroes.' That's all I could hope for."

—◆—

Another insider drew a different lesson from the story of the Challenger Seven. "It's about Reagan." Sipping coffee at a Dunkin' Donuts beside a highway, he says, "Please don't use my name. The fact is, the order to launch came down from the White House, and it had to do with Christa." He was privy to discussions at NASA and in the Reagan administration during and after the Teacher in Space program. "They'd had delay after delay, and now the president's about to give his State of the Union speech. He may even wave to Christa as the shuttle flies over DC. The story is, President Reagan got on the phone with the head of NASA. Reagan says, 'You've got schoolchildren all over the country watching and waiting.' So the message comes down: 'If you don't get that bird up in the air, we'll cut your funding.'"

Rumors that the president had ordered the launch began circulating in the first days after the accident. Richard Feynman looked into them "surreptitiously," as he put it. Feynman had heard that the president hoped to talk with Christa from orbit during his speech. The *New York Times* reported that "President Reagan's aides were in telephone contact with the space agency before the launching of the shuttle *Challenger*, but there was no evidence the communications entailed pressure on the launching schedule."

White House spokesman Larry Speakes collected twenty-eight sworn affidavits affirming "no outside pressure on NASA managers to launch *Challenger*." According to the *Times*, "Mr. Speakes said only calls from the White House had been studied, and left open the possibility that calls had come to the White House from NASA." Feynman suspected that the order to launch, if there was one, was never spelled out. "People in a big system like NASA *know* what has to be done *without* being told," he wrote.

"That's all you can prove," David Sugar added in 2020. "You won't find a paper trail." His meaning is clear: there is no paper trail. Sugar, an artist who helped the Challenger Seven engrave their names on the champagne bottle they planned to open when they returned to Earth, spent part of the night before the launch with the crew members' families. He wasn't surprised to hear of an internal NASA memo headed "Information on Presidential Phone Call Opportunities for Flight 51L," detailing what the memo called "Presidential availabilities to talk worldwide with Christa and the others." Everyone connected to the mission felt the pressure to launch.

"Cheryl McNair was really tense the night before," Sugar recalls. "Someone said, 'Don't worry. Do you think they'd let anything happen to the shuttle with the Teacher in Space on it?' Then they call off the launch because of the cold. Then we get the word: it's on again. We all knew what the president wanted. What the agency wanted. What the system wanted."

Today the space agency is "trying to evolve," Ciannilli says. After creating "Forever Remembered," he pitched KSC director Cabana another idea.

"I want to start a whole new program," he told Cabana. "Hear me out before you say no."

The program Ciannilli had in mind would delve into the agency's worst mistakes, the ones that killed astronauts. "We've had great successes in the sixty years of our space program. But we did some things wrong, and there are lessons in that." He wanted to discuss those lessons with agency employees.

"An agency-level program?" Cabana asked. "How would you pay for it?"

"I've got some ideas."

"If you can line up a budget for it, I'll back you."

NASA being NASA, Ciannilli's program had a mouthful of a title: the Apollo Challenger Columbia Lessons Learned Program, or ACCLLP. During one key moment in his initial talks to coworkers he described the teleconference on the night before *Challenger* launched. Ciannilli saw "no malice" in Mulloy and Kilminster, the managers who overruled their engineers. "It might be easier if there were bad guys, but you know what? People under pressure make mistakes." He asked his listeners to picture themselves in that teleconference at the end of a long workday that followed half a dozen delays, including the stripped-bolt debacle on Monday. "What would you do if your boss asked you, 'When are we gonna launch, next April?'" What if the program was falling behind schedule? What if arguing for another delay would cost you your job? What if the shuttle launched anyway and *nothing went wrong*?"

In his view it's easy to blame Mulloy and Kilminster—or Lucas or Reagan, for that matter. "That's what we've always done, isn't it? There's a terrible accident, and then there's a presidential commission, and then we go back to ops normal. It's like driving past a wreck on the highway. We see all the smoke and the ambulance, and after that we've got our hands at ten and two on the steering wheel. For a few miles. Twenty minutes later we're driving with one leg out the window and our toes on the wheel."

With new missions to the moon and Mars planned for the

2020s and beyond, his talks gained urgency. "NASA's not the only game in town anymore. The future of space is public *and* private." Addressing curious crowds at Richard Branson's Virgin Galactic, Jeff Bezos's Blue Origin, Elon Musk's SpaceX, and other companies, he urged them to learn from the agency's worst moments. Without constant diligence, he said, "Murphy's Law is gonna get you." Ciannilli feels a duty to get that idea across to anyone willing to listen. "This is how the mission continues. And I have a real feeling that the crews are helping me."

A key part of his message is, "The problem with *Challenger* wasn't the machine. The machine was trying to talk to us, but we didn't listen." Ciannilli believes the agency and the country have made progress since 1986. "We've done a lot for diversity when it comes to race, gender, and sexual orientation. We need more diversity of opinion. We need to build in a tolerance for people who will throw their careers in front of a runaway train. And when they do, we need bosses who will say, 'I've got your back.'"

—◆—

In 2018, aboard the International Space Station, Ricky Arnold and Joe Acaba performed what NASA called "Christa's Lost Lessons"—the experiments she had rehearsed. The space station, built by shuttle astronauts and cosmonauts starting in 1998, was the shuttle program's crowning achievement. Astronauts Arnold and Acaba had both taught middle and high school science before joining NASA; Arnold credited the Teacher in Space mission with inspiring him to become a teacher in the first place. "The whole country was invested in that mission," he said. Now he and Acaba rounded up Alka-Seltzer tablets, rubber balls, and other zero-G props "to bring that mission full circle."

Before filming Christa's lessons, Arnold and Acaba asked for Steve McAuliffe's blessing. Judge McAuliffe, still serving as a senior US District Court judge in New Hampshire, had no objection and no comment.

"Steve doesn't talk about Christa. Not publicly," says the *Monitor*'s Ray Duckler. The judge's silence was frustrating to a newsman like Duckler, "but you have to admire it. He put his family first." Over the years, Steve McAuliffe turned down the *Today* show, the *Oprah Winfrey Show*, book deals, TV and movie deals. He did not participate in a 2020 Netflix documentary on Christa's mission. Says Duckler, "He thought his family had a right to something like a normal life."

—◆—

In January 2020, June Scobee returned to Arlington National Cemetery for the thirty-fourth commemoration of her husband's burial. The cemetery's rows of white markers always gave her a chill, but there was happiness here, too, she said. It was at a 1988 sunrise service at Arlington National that she met US Army general Don Rodgers, who was mourning his wife. She and the general married a year later.

As June Scobee Rodgers she led the Challenger Centers from their beginnings at a single location in Houston. With help from the other *Challenger* families, astronauts, politicians, celebrities, and NASA administrators, her brainchild had gone through growth spurts and crises including a near-bankruptcy in 2008. That was when astronaut Bill Readdy joined the Challenger Centers' board of directors and helped put them on a sound financial footing. By the time June and other surviving family members gathered at Arlington in 2020, more than 5.5 million schoolchildren had "flown" or served at Mission Control at one of forty-three Challenger Centers in twenty-six states as well as Canada, England, and South Korea. Thousands more would go on remote-learning adventures as the Challenger Centers went online during the pandemic school year of 2020–21.

"June's kind, and charming—and tougher than nails," says one of her astronaut friends. "We call her the titanium magnolia."

Stepping off a NASA bus at Arlington National with Jane Smith, pilot Mike Smith's widow, on January 30, 2020, June carried a bouquet of roses in each arm. The two of them walked to the Challenger Memorial, where the unidentifiable remains of the crew were commingled and buried together in 1986. The memorial is a white stone with a brass plaque showing the smiling faces of Scobee, Smith, Judy Resnik, El Onizuka, Ron McNair, Greg Jarvis, and Christa McAuliffe. The opposite side of the stone is engraved with the poem "High Flight."

NASA deputy administrator Jim Morhard spoke for the agency that day. Mentioning each crew member in turn, he closed with "Christa McAuliffe, a social studies teacher at Concord High School in New Hampshire." In his next breath Morhard referred to the lessons Mike Ciannilli talks about. "I ask all in the NASA family to voice your opinions," he said, "to be unafraid to speak up, so that the guiding principle of our NASA culture remains: safety."

After a moment of silence, June placed one of her bouquets of roses at the foot of the *Challenger* marker. No one was there that day to lay flowers at the nearby memorial to the *Columbia* astronauts, but June Scobee Rodgers plans ahead. The second bouquet was for them.

Later, over lunch at a Washington restaurant, she wished Morhard had mentioned Christa's favorite quote: "I touch the future—I teach." June remembered Christa's telling her why she was willing to leave her family for six months and ride a rocket into space.

"She wanted to show that teachers have the right stuff, too. That's why she did it." She said Christa would have been proud of the many young people she inspired to become

252 ≡ KEVIN COOK

schoolteachers—more than a dozen from Christa's Concord High classes, and hundreds if not thousands more all over the country and the world. They included Christa's daughter, Caroline, who became a grade-school teacher.

June had just heard from Caroline's dad, who objected to a line in the Challenger Centers' bylaws that described the crew as "astronauts." Steve knew that NASA had posthumously promoted payload specialist McAuliffe to astronaut, but he wanted to strike that line. "Christa did *not* want to be called an astronaut," he said. "She was always a teacher, and proud of it." He and June changed the Challenger Center bylaws, which now refer to "the astronauts and the first Teacher in Space."

Christa's legacy mattered deeply to Steve, though he didn't talk about it publicly. "One young woman told me she followed Christa's example and enthusiastically became a teacher," he recalls today. "This teacher was committed and dedicated, but she was frustrated by out-of-control classrooms, irritated by inadequate support, overwhelmed by her all-day and most-of-the-evening schedule, thinking about finding a different career. She was glad to hear that Christa and most of her teacher friends went down that very same road when they started out. I was pleased to repeat what I'd heard Christa tell other young teachers: that she cried almost every night from September through March her first year. But it's worth persevering because teaching is a noble calling and a daily joy."

Judge McAuliffe turned seventy-two in 2020. It had been fifty years since he married his high school sweetheart at St. Jeremiah's Church in Framingham. He had reached the age she anticipated during her training, when she talked about "thirty-six exciting years behind me and thirty-six more ahead." Had she lived, he says, "I think she would be in the forefront of an effort to teach young people civics, history, and critical thinking." He thinks

Americans could use "a remedial course in basic civics. That's something she always stressed in her teaching. You can't expect citizens to value and protect our form of government—the separation of powers between three branches of government, the fact that the military must always be under civilian control—if they barely understand it."

One thing he found "noteworthy in this year of COVID-19" was a new appreciation for teachers. "Admiration for them soared as parents and grandparents tried to do the job with their children at home. Most of them weren't very good at it. Why? Because teaching is hard. It requires education, training, aptitude, skill, and hard, hard work. I hope this global experience we've all shared in 2020 and 2021 lifts teachers in the eyes of the public."

How should people remember Christa McAuliffe?

"As an ordinary person who embraced an extraordinary opportunity," he says. "As a skilled classroom teacher who loved her students and her work. As a woman who fiercely loved her family, loved and valued her friends, and was loved in return. As a happy woman who laughed often and lived a life guided by optimism and a fearless pursuit of her dreams. As a person who brought joy and inspiration to those fortunate enough to know her, and inspired people who only knew *of* her."

A NOTE ON SOURCES

While working on *The Burning Blue* I relied on hundreds of sources including interviews in person, by phone, and by email with scientists, astronauts, crew members' families, and NASA personnel past and present, as well as NASA documents, newspaper and magazine accounts, videos, TV transcripts, and other records dating back to the dawn of the space program. I learned something from everyone I met and every document I read. NASA's Historical Reference Collection at the agency's headquarters in Washington, DC, and the Johnson Space Center's Oral History Project were irreplaceable resources. I often turned to six books: Grace Corrigan's *A Journal for Christa* (Lincoln: University of Nebraska Press, 1993); Robert Hohler's *"I Touch the Future": The Story of Christa McAuliffe* (New York: Random House, 1986); Diane Vaughan's *The Challenger Launch Decision* (Chicago: University of Chicago Press, 1996), a study of events leading to the launch; Allan McDonald and James R. Hansen's *Truth, Lies, and O-Rings* (Gainesville: University Press of Florida, 2009), an account of McDonald's work at Morton Thiokol and the teleconference on the eve of the launch; Kerry Joels's *The Space Shuttle Operator's Manual* (New York: Ballantine Books, 1982), a guide to everything about the shuttle; and astronaut Mike Mullane's *Riding Rockets* (New York: Scribner, 2006), the most entertaining space book since *The Right Stuff*. I also relied on the 1986 "Report of the Presidential Commission on the Space Shuttle Challenger Accident" (available at https://history.nasa.gov/rogersrep/genindex.htm), better known as the Rogers Commission Report, and NASA's 1,139-page "Shuttle Crew Operations Manual" (available at https://www.nasa.gov/centers/johnson/pdf/390651main_shuttle_crew _operations_manual.pdf), compiled originally by astronaut Jim Wetherbee from his notebooks and those of astronaut Michael Smith.

A note on tenses: If a source spoke to me during my work on the book, I put her or his quote in the present tense. Otherwise I use past tense.

Prologue

Dialogue between the shuttle, Launch Control, and Mission Control comes from NASA transcripts. Christa McAuliffe's description of a prelaunch delay ("Put on a motorcycle helmet . . .") is from a communication between her and Concord High School teacher Eileen O'Hara reported in the school newspaper, the *Crimson Review*.

1

McAuliffe recalled watching Alan Shepard's 1961 flight in her application for the Teacher in Space program. Details of her youth come from the Christa McAuliffe Collection at Framingham State University; her mother Grace's *A Journal for Christa*; Hohler's *"I Touch the Future"*; and interviews with Christa's friends Mary Liscombe, Pat Berlandi, Susan DuPonte Conway, and Michael Conway, who shared memories of their college days together. Hohler cited the journal of the pioneer woman who rode horseback with her baby. Details of Christa's wedding and honeymoon are from *A Journal for Christa* and a letter to her mother. John Noble Wilford quoted Christa's calling Concord "a Norman Rockwell kind of place" in the *New York Times* on January 5, 1986. Eileen O'Hara remembered her friend Christa in email exchanges with me. The *Monitor* featured President Reagan's announcement of the Teacher in Space program in its August 28, 1984, edition. My account of her applying for the Teacher in Space program comes from her correspondence, the *Monitor*, the Framingham-based *Middlesex News*, her mother's and Hohler's books, and her application for the program. Her reaction to her husband's job offer from the Justice Department comes from Steve McAuliffe, quoted in the *Boston Globe*. The furnishings and decor of her home come from her mother's book and contemporary photos. Her activism as a teachers' union leader was recounted in the *Monitor*. Hohler reported that the Bow Memorial administration "wasn't ready for a woman yet." Cindy Edson, Kevin Swopes, and Andrew Cagle recalled what it was like to be at Concord High in the year of the Teacher in Space. Grace Corrigan's book told the mildly scandalous tale of Christa's joining other moms to score discounts by posing as grocery-store buyers. Hohler noted that Caroline's name was a tribute to President Kennedy's daughter, which Dr. June Scobee Rodgers confirmed to me; *A Journal for Christa* noted that the name also honored Christa's Aunt Carrie.

2

Christa's mother recalled the troubled student who turned up on the McAuliffes' doorstep on the night before Christa flew to Washington. Teacher in Space semifinalists Bob Veilleux and Richard Methia shared their recollections with me. Mike Pride, then editor of the *Monitor*, shared with me his impressions of Christa and the impact her adventure had on Concord. Her rise from semifinalist to Teacher in Space was chronicled by Hohler and Grace Corrigan, and soon by many newspaper and TV reporters. Astronaut David Hilmers helped me understand the Rescue Sphere. Hohler recounted the Teacher in Space finalists' oxygen-deprivation and

Vomit Comet tests. Methia shared memories of the finalists' experiences with me, including the Lunar Odyssey ride that killed Todd Walker. Kathleen Beres recalled Christa's reaction to Walker's accident in a February 9, 1986, *Chicago Tribune* story. Reagan's speech to the finalists, Christa's selection as the Teacher in Space, and her moment with Vice President Bush are available on YouTube and are well worth watching.

3

Grace Corrigan recounted her July 1985 car trouble in her *Journal for Christa*. *Monitor* editor Mike Pride, who went on to oversee the Pulitzer Prizes, and *Monitor* writers Bob Hohler, Ray Duckler, and Ralph Jimenez helped me picture Concord in 1985–86. Hohler cited Wally Schirra's description of the shuttle as a thing "we're still learning to fly." Christa's TV appearances and magazine interviews were widely reported. Her letters home come from her mother's book and the McAuliffe Collection at Framingham State University. Grace Corrigan's book described President Reagan's quip (*"Watch this"*). Hohler reported her comments about the White House dinner where she and Steve met Raquel Welch and Sylvester Stallone. Hohler and the *Washington Post* described her work with her JSC tutor Bob Mayfield.

4

Christa's rehearsals for her space lessons were recorded by NASA. They can be seen—along with astronaut Ricky Arnold's re-creation of "Christa's Lost Lessons" on the International Space Station in 2018—at challenger.org. My account of what she encountered at JSC comes from interviews with Dr. June Scobee Rodgers, Cheryl McNair, and current and past NASA workers as well as news accounts, the McAuliffe Collection at Framingham State, her mother's book, Mullane's *Riding Rockets*, and JSC's in-house newspaper *Space News Roundup*. John Noble Wilford reported her worry about falling off a treadmill in the *New York Times* on January 5, 1986. Multiple sources reported her quotes and experiences as she became a celebrity. Hohler described her moments at home with her family. Frank Hughes, NASA's former chief of spaceflight training, helped me understand how she and her crewmates trained for STS-51L. Bobby Mayfield recalled Christa in an oral history interview for the Veterans History Project, held at the Library of Congress. The *Los Angeles Times* reported on their clashes; Hohler told of Mayfield's praise for her. Christa's letters at Framingham State, her mother's book, and my conversations with Dr. June Scobee Rodgers added details of her relationship with Judith Resnik. Hoot Gibson's joke about Congressman Bill Nelson's hoping to see angels is from Gibson's interview for JSC's Oral History Project.

5

Christa's correspondence with Eileen O'Hara is in the McAuliffe Collection at Framingham State. O'Hara also corresponded with me, confirming details and adding recollections of her friend Christa. My account of the crew's gatherings in Clear

Lake is drawn from conversations with Dr. Rodgers, Cheryl McNair, and other crew members' families as well as the JSC Oral History Project. Resnik's former husband, Mike Oldak, provided invaluable insight into her life and character. Dick Scobee's life story comes from Dr. Rodgers, Kathie Scobee Fulgham, several astronauts, and Frank Hughes, chief of training at JSC. The staff of the *Washington Post* reported helpful details in the *Post* and the 1986 book *Challengers* (New York: Pocket Books, 1986).

6

Hohler's account of Mildred Doran and the 1927 Dole Air Race led me to contemporary news reports and a March 2011 story by Richard A. Durose in *Air & Space* magazine. Grace Corrigan recalled her daughter's line about "all those rockets and fuel tanks." The *Monitor* and the *New York Times* reported on Steve's career and his trip to the Supreme Court. My portrayal of Christa's training draws from press reports and other sources including NASA video recordings, the JSC Oral History Project, Mullane's book, and Frank Hughes's vivid recollections. Astronaut Jim Wetherbee told me about the stray bolt and cookie crumb he encountered on the flight deck. The *Chicago Tribune* reported Steve Cunningham's views of the training he and Greg Jarvis went through at JSC. Hohler described Christa's clash with her NASA editors over the teachers' guide she refused to endorse. Her letters to Eileen O'Hara described parts of her training. The *Monitor* reported William Bennett's visit to Concord High. Kevin Swope, a Concord High student at the time, recalled Bennett's visit and many other details of life at CHS in 1985–86. Grace Corrigan recalled serving as her daughter's substitute at Framingham State's homecoming in 1985. The card Christa sent her ("Behind every great woman . . .") is in the McAuliffe Collection at Framingham State. The script for her first "Lesson from Space" is from a NASA memo in the agency's Historical Reference Collection. My depiction of the good-natured trick Scobee and the crew played on her on the day of their official photo owes details to Eileen O'Hara, Hohler, the *Washington Post*, and the *New York Times*.

7

Hohler and many others covered training for *Challenger*'s mission. Dr. Rodgers recalled her husband reminding Christa to "be serious" about it. Slate.com reported on NASA's Safety Advisory Panel's study of space-shuttle ejection seats. Astronaut Mike Coats recalled long simulations in his JSC Oral History Project interview. Mullane told of Guion Bluford's "medical emergency" in *Riding Rockets*. Frank Hughes spoke to me at length about training lore and practice. NBC News's Charles Lam told the story of JSC sims and Ellison Onizuka's scuba gear in a January 25, 2016, profile of Onizuka. The story of toilet trouble on a shuttle flight is from the STS-8 Flight Crew Report at JSC. My account of Louie Cartier's hostage-taking at Concord High School draws from reporting by the *Monitor*, Hohler, United Press International, a fine 2018 story by Natalia Megas for narrative.com, and my talks with Concord High students and teachers. A Concord High student who asks not

to be identified told me that it was a member of the football team who had bul-
lied Cartier. Ralph Jimenez of the *Monitor* helped me put the Cartier story in con-
text. Details of Dr. Judith Resnik's life come from my conversations with Michael
Oldak, Dr. Rodgers, astronaut Norm Thagard, and Frank Hughes, plus a moment
with Charles Resnik. Other sources for Resnik's story include JSC's Oral History
Project, *Judith Resnik, Challenger Astronaut*, by Joanne E. Bernstein and Rose Blue
(New York: Lodestar Books, 1990); Amy E. Foster's *Integrating Women into the
Astronaut Corps* (Baltimore: Johns Hopkins University Press, 2011); and a fine pro-
file in the winter 1984 issue of *Carnegie Mellon Magazine*. The *Washington Post's*
Challengers quoted her mentor Robert Newcomb on her "turmoil" and divorce.
Michael Collins told his off-color polar bear story in his *Carrying the Fire* (Lan-
ham, MD: Cooper Square Press, 1974). The supermarket tabloid *The Globe* head-
lined the first female astronauts as "Glamournauts, six NASA lovelies" on March
18, 1981. Resnik's TV talks with NBC's Tom Brokaw are on YouTube. Astronauts
Hank Hartsfield, Coats, and Mullane recalled her duel with the IMAX camera on
Discovery.

<h1 style="text-align:center">8</h1>

Hohler described the December 1985 press conference that began with a mention
of "Christa McCoffee." *USA Today*, the *Houston Post*, and *Space News Roundup*
reported on it. Launch director Gene Thomas, among others, described *Challenger's* crew as the most diverse in NASA history in his memoir *Some Trust in Chariots* (Maitland, FL: Xulon Press, 2006), a book that also provided details of life in
and around the Firing Room, including an anonymous engineer's parody of "The
Super Bowl Shuffle." My account of Dr. Ron McNair's story comes from conversations with Cheryl McNair, supplemented by Carl McNair's *In the Spirit of Ronald
E. McNair, Astronaut* by Carl S. McNair (Atlanta: MAP Publishing, 2011) and Jon
Kirby's fine November 19, 2019, McNair story in the *Oxford American*. McNair's
smashing appearance in *Scientific American* came in the April 1979 issue. Michael
Cassutt's *The Astronaut Maker* (Chicago: Chicago Review Press, 2018) recalled
George Abbey's considering Guion Bluford and McNair equally qualified, "with
McNair cited for his outstanding public presence" while Bluford "was a pilot and
combat veteran." Dr. Rodgers, whose heartfelt book *Silver Linings* (Macon, GA:
Smyth & Helwys Publishing, 1995) added to our conversations about the *Challenger* crew, told me about the circumstances surrounding Christa's return to Concord for the holidays in 1985. Hohler reported on her visit, which lasted through
First Night and New Year's Day. *New Hampshire* magazine described her First
Night duties on April 28, 2011. The AP and UPI wire services and New England
newspapers reported the Patriots' victory in the 1985 playoffs and their fans' cele-
brations. Former Patriots star Brian Holloway recalled that day and his postgame
moment with Christa in one of our conversations. Malcolm McConnell's *Challenger: A Major Malfunction* (New York: Doubleday, 1986) detailed flow director
Harrington's sending a T-38 to pull spare parts from *Columbia* and Mike Smith's
worries about the weather. Astronaut Bob Crippen described the jet's "travel pod"

in his JSC Oral History interview. Cheryl McNair and Dr. Rodgers described their flight from JSC and the moments that followed in their talks with me. Steve Cunningham, Greg Jarvis's backup as payload specialist from Hughes Aircraft, recalled Jarvis's comment on seeing the stack on its launchpad in the February 12, 1986, *Chicago Tribune*.

9

John Logsdon, founder of the Space Policy Institute at George Washington University and perhaps the premier authority on the history of the US space program, helped me understand the dynamics that drove the shuttle program. The shuttle specs I cite come from NASA archives. Scobee and other shuttle commanders compared their work to "flying a brick." Former orbiter processing chief Terry White confirmed that and other details to me. Information on NASA's budget and workforce after the 1969 workforce come from publicly available records and agency archives. An astronaut who prefers not to be named described to me other aspects of agency life in those days. NASA's 1983 marketing brochure is in the Historical Collection at NASA headquarters. The NASA document on developing the Teacher in Space lessons, "Hardware Development for Teacher in Space Activities," is in the Historical Reference Collection at NASA headquarters. Dr. Rodgers described the crew's time at the beach house in our conversations; Grace Corrigan's *A Journal for Christa* contributed details. Mullane's *Riding Rockets* was the most colorful of several renditions of Resnik's experiences on *Discovery* in 1984, with Hartsfield's JSC Oral History Project interview not far behind. Scobee recalled the bees on his first mission in his wife's *Silver Linings*; she elaborated in our talks. Hartsfield and Mullane recalled the invisible fire that threatened their mission. Gregg Easterbrook's prescient story on space-shuttle dangers appeared in the April 1980 issue of *Washington Monthly*. Hohler told the story of Scott McAuliffe's flight from New Hampshire to Florida. His book and Grace Corrigan's described the Saturday-evening party at the Holiday Inn in Orlando.

10

NASA's requirements for crew members' personal preference kits are detailed in several documents in the agency's Historical Reference Collection. Grace Corrigan recalled the contents of her daughter's PPK. Hohler and Colin Burgess's *Teacher in Space* (Lincoln: University of Nebraska Press, 2000) contributed others. McConnell, the *Washington Post*'s *Challengers*, and other sources including the *Orlando Sentinel*'s "The Final Hours of Space Shuttle Challenger" (January 28, 2006) reported on pilot Smith's concerns about the weather. My account of Super Bowl Sunday draws from contemporary media, Hohler, Grace Corrigan, and interviews with Dr. Rodgers, Kathie Scobee Fulgham, Cheryl McNair, and David Sugar. *Air & Space* magazine (September 2010) described the card game Possum's Fargo, which several astronauts told me about. Hoot Gibson recalled the sights and sounds of the launchpad in his JSC Oral History Project interview. My account of the scrub on Monday, January 27, relies on dozens of sources new and old. McConnell described the moment between Scobee and Corlew after the scrub. Launch director Thomas recalled "The Super Bowl Shuffle" parody in his memoir. Mullane recalled his last conversation with Resnik. Barbara Morgan remembered Christa's writing recommendations for students on the eve of the flight in a 1996 essay for *USA Today*. David Sugar quoted his friend McNair ("The launch is back on") in a talk with me. Grace Corrigan quoted her daughter's assurance that "we're going tomorrow" in her book and in a CNN interview that appeared on January 22, 2006.

11

Dr. Rodgers recalled her launch-morning conversation with her husband in her book *Silver Linings* and conversations with me. Flight director Brock "Randy" Stone told of his phone call with Chuck Knarr in his JSC Oral History Project interview. Details of the morning's events come from the books cited above; press accounts including the *Sentinel*'s fine "The Final Hours of Space Shuttle Challenger" (January 28, 2006), NASA video footage; and my interviews with astronauts, agency engineers, and technicians. McConnell's *Challenger* told the story of the duplicate cake and Onizuka's warm jacket. Hohler, Grace Corrigan, McConnell, Dr. Rodgers, Sugar, the *Sentinel*, the *Monitor*, CNN, *Popular Mechanics*' February 2016 "Oral History of the Space Shuttle *Challenger* Disaster," and the Rogers Commission Report filled out my account of the hours before launch. Teacher in Space semifinalist Bob Veilleux helped me with photos and recollections of that morning. Grace Corrigan recalled, "The word was out that today was the day" in *A Journal for Christa*. Johnny Corlew's role in launch preparations derives from NASA videos, *Popular Mechanics*' "Oral History of the Space Shuttle Challenger Disaster," the *Washington Post* staff's *Challengers*, CNN, and other sources. The crew's communications with Launch Control come from NASA transcripts. McConnell portrayed the roles of Rockwell executives Rocco Petrone and Bob Glaysher. Prelaunch chatter between Scobee and his crew comes from NASA transcripts. My description of the middeck owes to Joels's *The Space Shuttle Operator's Manual* and dozens of other sources. Christa's rehearsals for her space lessons can be seen at challenger .org/challenger_lessons/christas-lost-lessons/. Grace Corrigan remembered worrying about the ice and telling her husband "she wouldn't come" if he tried to pull Christa off the shuttle. Dr. Rodgers described the scene on the roof of the Launch Control Building to me. Eileen O'Hara, Kevin Swope, and the *Monitor*'s coverage helped me describe the scene at Concord High School. Agency records in NASA's Historical Reference Collection tallied the temperatures of shuttle launches. Hoot Gibson described the thunder of launch in his JSC Oral History Project interview. Sally Ride did the same in her JSC Oral History talk.

12

My description of the launch and the mission's first seventy-three seconds owes to NASA's records including video documentation as well as the books and press accounts cited above. Dr. Rodgers described the scene on the roof of the Launch

Control Building for me. Space-program historians Logsdon and Bill Barry helped me understand the events of that morning. Some important details of my account are from astronaut Rhea Seddon's *Go for Orbit* (Murfreesboro, TN: Your Space Press, 2015). Grace Corrigan remembered her son-in-law's anguish and George Abbey's retreat. The *Monitor* covered the scene at Concord High, which Cindy Edson, Kevin Swope, and Andrew Cagle, all students at the time, recalled for me. Hohler described the librarian ringing church bells and the policeman kneeling in the snow. My description of President Reagan's reaction to the disaster owes details to Logsdon, contemporary press reports, and speechwriter Peggy Noonan's *What I Saw at the Revolution* (New York: Random House, 1990) as well as her "Challenger Disaster: The View from the Oval Office" in the December 2015 issue of *AARP* magazine. Dr. Rodgers and Cheryl McNair recalled the scene in the Launch Control Building in talks with me. My quotes from TV news anchors Tom Brokaw, Peter Jennings, and Dan Rather are from their broadcasts on January 28, 1986. Reagan's speech can be seen on YouTube or at reaganfoundation.org. The Ronald Reagan Presidential Library provided vital information including the president's diary entries for January 28 ("A day we'll remember for the rest of our lives") and other days involving *Challenger*. *People* magazine put Christa on its February 10, 1986, cover.

<h2 style="text-align:center">13</h2>

There are no indications that any crew member died when the orbiter broke up seventy-three seconds after launch. In NASA's official postmortem, submitted to Associate Administrator for Space Flight Richard Truly six months to the day after the accident (available online and in NASA's Historical Reference Collection), astronaut Joseph Kerwin, an MD serving as biomedical specialist at JSC, found that the still-intact crew cabin had been so damaged by impact with the ocean that there was little he could say for certain. Kerwin estimated that G forces rose from twelve to twenty Gs immediately after the breakup. "These accelerations were quite brief. In two seconds, they were below four Gs; in less than ten seconds, the crew compartment was essentially in free fall," he wrote. "Medical analysis indicates that these accelerations are survivable, and that the probability of major injury to crew members is low." He reported that four of the crew's seven PEAPs had been recovered; three had been activated. (Scobee's had not been activated.) As for depressurization, "Impact damage was so severe that no positive evidence for or against in-flight pressure loss could be found," though the fact that at least three PEAPs were turned on suggest that depressurization, if it happened, wasn't sudden. In conclusion Kerwin wrote that "the skilled and dedicated efforts of the team from the Armed Forces Institute of Pathology, and their expert consultants, could not determine whether in-flight lack of oxygen occurred, nor could they determine the cause of death." My account of the minutes between breakup and impact with the ocean draws from documents in the Historical Reference Collection as well as the Rogers Commission Report; Mullane's *Riding Rockets*; the JSC and *Popular Science* oral histories; my talks with Hilmers, Thagard, and an astronaut who asked not to be named, as well as Hughes, Joels, Logsdon; and virtual visits to the Crew Compartment Trainer now at the National Museum of the United States Air Force at Wright-Patterson Air Force Base in Ohio. Seddon's *Go for Orbit* mentions switches thrown after the breakup. Carl McNair's *In the Spirit of Ronald E. McNair, Astronaut*, Jay Barbree's report for NBC News ("The Challenger Saga: An American Space Tragedy," January 20, 2004), and *Aviation Week and Space Technology* (August 4, 1986) added important details. Hughes told me why he thinks it is more likely that Onizuka rather than Resnik activated Smith's PEAP. Jim Wetherbee provided the detail that mission specialist 1 (Onizuka) would activate the pilot's PEAP during training. John Young recalled seeing the cabin fall in his *Forever Young* (Gainesville: University Press of Florida, 2012). Launch director Thomas recalled his hope that Scobee might save the shuttle in his *Some Trust in Chariots*. Astronaut Bob Overmyer told NBC-TV that Scobee tried to fly his craft "all the way down." Several press accounts, including Dennis Powell's influential "Obviously a Major Malfunction" in *Tropic*, the *Miami Herald*'s Sunday magazine (November 13, 1988), quoted an investigator saying the crew "could have swam home," a line I turned to "swum." Jarvis's calling the stack "a keg of dynamite" and Ed Corrigan's saying Jarvis and Christa may have been "holding hands" come from the February 10, 1986, *New York Times*. Jeff Hanley's recollections are from his JSC Oral History Project interview.

14

McDonald recalled the scene in the Launch Control Center after *Challenger* broke up. Press reports, prominently including AP and the *New York Times* of December 18, 1996, as well as my conversations with present and former NASA workers, including Terry White and Jill Vogel, confirmed that shuttle debris has often washed up on Florida beaches and still occasionally does. Press accounts and government documents describe the navy's salvage operation as the largest in history, including the navy's own "Space Shuttle Challenger Salvage Report" of April 29, 1988. Frank Hughes remembered his reactions in a conversation with me. Doc Pepper's remarks come from an email exchange with me. The document noting Senator Patrick Leahy's request about "shuttle launch delays" can be found in NASA's Historical Reference Collection. Johnny Corlew recalled taking leave after the event in *Popular Science*'s 2016 oral history. Grace Corrigan mentioned Christa's recommendation to Matt Mead in *A Journal for Christa*. Then *Monitor* editor Mike Pride described Concord after the disaster in exchanges with me. Grace Corrigan described the *National Enquirer* reporter on her doorstep. Dr. Rodgers told me about the days after the accident. Several sources quoted post-accident jokes to me; those and other jokes can easily be found online. Michael Collins's "Riding the Beast" appeared in the *Washington Post* on January 30, 1986. Art Buchwald's column ("I'm a teacher too") ran nationwide through the *Los Angeles Times* syndicate on February 7, 1986. Speechwriter Peggy Noonan recalled President Reagan's work habits in her memoir *What I Saw at the Revolution*. Dr. Rodgers told me about her moment with the president before his speech at JSC. Grace Corrigan mentioned her moment with Nancy Reagan in her memoir. Reagan's speech is on YouTube; I found useful background on his speech at the Reagan Library (reaganfoundation.org). Grace Corrigan recalled Christa's parents' moments with Caroline Kennedy and John F. Kennedy Jr.

15

Dale Uhler's comments on the salvage operation come from documents in NASA's Historical Reference collection and contemporary news accounts. The search for Object D was described in the Rogers Commission Report. The *New York Times* story "NASA Had Warning of a Disaster Risk Posed by Booster" appeared on February 9, 1986. McDonald discusses the so-called Apocalypse Letter in *Truth, Lies, and O-Rings*. Grace Corrigan described her son Kit's and husband Ed's reactions after the disaster. The *Chicago Tribune* reported Steve McAuliffe's reaction ("Steve feels burned") on January 25, 1987. McDonald and numerous other sources told of the graffito calling Morton Thiokol "murderers." McDonald and Richard Lewis's *Challenger: The Final Voyage* (New York: Columbia University Press, 1988) detailed Thiokol history. McDonald told of mishaps at Thiokol's plant in Utah. Morton Thiokol's corporate records are held at the Hagley Museum and Library in Wilmington, Delaware. AP reported on the salvage effort and diver Thomas Stock's comments on April 14, 1986. The insider who described Resnik's body's being outside the crew cabin is a prominent former NASA employee. Dennis Powell's 1988 *Miami Herald* report, the *Washington Post*, and ABC News, contributed to my account. Mullane recounted John Young's reaction. Astronaut Rhea Seddon described her examination of the crew cabin's debris in *Go for Orbit*. Grace Corrigan remembered her daughter's funeral in her memoir.

16

Richard Feynman related his phone call from William Graham and his Rogers Commission work in his memoir *"What Do You Care What Other People Think?"* (New York: W. W. Norton, 1988). Other details of Feynman's life and commission experiences come from James Gleick's *Genius: The Life and Science of Richard Feynman* (New York: Pantheon Books, 1992). My description of the commission's proceedings owes to the Rogers Commission Report and transcripts and videotapes of its hearings. The *New York Times'* David Sanger told the producers of a 2020 Netflix documentary, *Challenger: The Final Flight*, that President Reagan told William Rogers not to "embarrass NASA." Feynman recalled a similar quote. The *Times* reported NASA's O-ring problems on February 9, 1986. The *Popular Mechanics* February 2016 oral history held General Kutyna's admission that the NASA document on O-rings' resilience came from Sally Ride. McDonald and Feynman recalled McDonald's interrupting a closed hearing to say that Thiokol's engineers had "recommended not to launch." My account of the January 27 teleconference draws from numerous sources including the Rogers Commission Report and books and press accounts cited above. Brian Welch's memo quoting Richard Truly's memo ("This won't be a namby pamby program") can be found in NASA's Historical Reference Collection. Dr. Rodgers recalled the day the Rogers Commission Report was released and Rich Scobee's reaction in a talk with me.

17

AP reported on the "Return to Flight" launch on September 29, 1988. The NASA memo on agency officials' potential liability was dated May 14, 1986. The agency reported Fairglen Elementary School's one-thousand-dollar donation in NASA news release 134-86, dated October 24, 1986. Astronaut David Hilmers described that flight to me by phone and in emails. Commander Rick Hauck described it in one of his JSC Oral History Project interviews. NASA transcripts added comments from *Discovery* astronauts. United Press International reported the interment of *Challenger* debris in missile silos on November 26, 1986. NASA photos documented the event. A NASA worker described the weeds around the site to me. KSC director Bob Cabana's quote comes from a NASA video posted on October 22, 2020. My description of the lawsuits that followed the accident rely on press accounts and my own talks with participants. Jane Smith's charges against NASA can be found in the Historical Reference Collection. Michael Oldak told me about his work for the Resnik family. Christa's grave is on a hill at Calvary Cemetery in Concord. Joe Dwinell shared his memories of that time with me. His recollections of Ed Corrigan added to what I read in Grace's book and her notes to Christa in the archives at Framingham State. Dr. Rodgers told me about her post-*Challenger* days and the Challenger Centers she and other family members built. She and others, including James Rosebush, who served in the White House as Nancy Reagan's chief of staff, helped me describe the Challenger Centers' early days. Brian Holloway told me about selling his Super Bowl ring to support the Challenger Centers. CNN (January 28, 1996) and the *Orlando Sentinel* (January 8, 2003) reported on Bruce Jarvis's leaving thank-you cards to drivers with the *Challenger* license plate. *Florida Today* covered the Space Mirror's 1997 mishap. Dr. Rodgers told me about the Challenger Centers' early days and her meeting with Ross Perot. Rick Hutto shared his recollections of *The Oprah Winfrey Show* with me. Joels recalled Lorna Onizuka's gesture on leaving a Challenger Centers fundraiser. John Denver's "Flying for Me" appeared on his 1986 album *One World*. Dr. Rodgers described her son's flyover of the 1996 Super Bowl to me.

18

The August 26, 2003, report of NASA's Columbia Accident Investigation Board confirmed that foam shedding was a long-standing problem the agency deemed "acceptable." A NASA report released on December 30, 2008, detailed the *Columbia* crew's last moments. Kathie Scobee Fulgham's letter to the *Columbia* children was reported by ABC News, the *Orlando Sentinel*, NASA.gov, and other outlets. NASA's Mike Ciannilli, head of the Apollo Challenger Columbia Lessons Learned Program (ACCLLP), shared his story in in-person, phone, and email exchanges with me. Several minor details of the missile silos holding *Challenger* debris came not from him but from present and former NASA workers who asked not to be identified. The person who spoke with me at a Dunkin' Donuts, who also asked not to be identified, had firsthand knowledge of the launch delays and NASA contacts who described growing pressure to launch the shuttle. Logsdon evaluated White House pressure "to launch *Challenger* no later than January 28 because President Reagan intended to mention the launch in his State of the Union address that evening" in his authoritative *Ronald Reagan and the Space Frontier* (London: Palgrave Macmillan, 2018). Feynman's accounts and my talks with David Sugar added to my account. I attended the January 2020 memorial at Arlington National Cemetery and spoke with Dr. Rodgers afterward. Judge McAuliffe graciously answered my questions in December 2020.

Other sources for details in the book include Michael Cassutt's *The Astronaut Maker* (Chicago: Chicago Review Press, 2018); Henry S. F. Cooper Jr.'s *Before Lift-Off* (Baltimore: Johns Hopkins University Press, 1987); Richard S. Lewis's *Challenger: The Final Voyage* (New York: Columbia University Press, 1988); David Hilmers's memoir *Man on a Mission* (Grand Rapids, MI: Zonderkidz, 2013); Joseph Trento's *Prescription for Disaster* (New York: Crown Publishers, 1987); *Spacefarers*, edited by Michael Neufeld (Washington, DC: Smithsonian Institution Scholarly Press, 2013); Ben Evans's *Space Shuttle Challenger* (Chichester, UK: Praxis Publishing, 2007); and Karen Bush Gibson's *Women in Space* (Chicago: Chicago Review Press, 2014).

ACKNOWLEDGMENTS

Thanks above all to Dr. June Scobee Rodgers, who has done more than anyone else to celebrate the lives and legacies of the Challenger Seven. You can't meet Dr. Rodgers without being stirred by her sunny optimism, intellect, and grit. I was lucky enough to spend several months in June's orbit before the COVID-19 pandemic and to stay in touch with her afterward. It was and is a privilege to know the Challenger Centers' titanium magnolia.

Kathie Scobee Fulgham welcomed me to the Astronauts Memorial Foundation at KSC. I'm an admirer of Kathie's thoughtful spirit and her writing, including her eloquent 2003 letter to the *Columbia* families. She put me in touch with Cheryl McNair and Mike Ciannilli, who made key contributions. Many thanks to Kathie, Cheryl, and Mike. I'm also grateful to Michael Oldak for sharing memories of his marriage to astronaut Judith Resnik, and to others I met at NASA's Day of Remembrance ceremonies at Arlington National Cemetery in January 2020, including Jane Smith and Chuck Resnik.

Eileen O'Hara helped Christa with her Teacher in Space application and subbed for her at Concord High in 1985–86. Thirty-five years later she shared memories of her friend with me. Thank you, Eileen. Concord High graduates Kevin Swope, Cindy Edson, and Andrew Cagle told me about Mrs. McAuliffe and their eventful senior year. Bob Veilleaux, who joined Christa in Washington as one of New Hampshire's candidates, and Teacher in Space finalist Rich Methia helped me understand the pleasures and pressures of what Christa went through.

Kerry Joels, coauthor of *The Space Shuttle Operator's Manual*, spent hours bringing me up to speed on shuttle history and the stack's machinery. (In that as all else in the book, any mistakes are mine.) I owe Kerry a ballpark beer at a Tulsa Drillers game.

NASA's Bill Barry, the agency's historian, aided my work before and after he retired in 2020. His staff, including Liz Suckow and her colleague Colin Fries at NASA headquarters in Washington, helped me find my way around the Historical Reference Collection. In Houston, expert interviewer Jennifer Ross-Nazzal helped guide me through the JSC Oral History Project's trove of Q&As. Frank Hughes, NASA's longtime chief of spaceflight training, was a source of invaluable details and great conversation. Current and former NASA employees Bert Ulrich, Connie Moore, Kamran Bahrami, Terry White, Doc Pepper, Lee Solid, Jill Vogel, and Tiffany Fairley helped me research parts of the book. Mary French at the George H. W. Bush Presidential Library and Museum helped me track down photos, as did Steve Branch and Jennifer Torres at the Reagan Presidential Library.

Five-time shuttle commander Jim Wetherbee's *Controlling Risk in a Dangerous World* helped me understand how a commander thinks. Wetherbee spent hours on the phone with me—I now know about 1 percent of what he knows about risk and switchology—and shared memories of Judith Resnik and Mike Smith. Astronaut David Hilmers helped me describe 1988's Return to Space mission. Astronauts Eileen Collins, the first female shuttle commander, and Tom Jones explained aspects of spaceflight. I'm also indebted to an astronaut and several NASA connections who asked not to be identified.

At Framingham State University, Mary Liscomb described her friendship with Christa and introduced me to their mutual friends Pat Berlandi, Susan Conway, and Michael Conway. Colleen Previte and her staff at the Whittemore Library guided me through Framingham State's Christa Corrigan McAuliffe Collection. I'm also indebted to Framingham's Judith Kalaora, whose heartfelt and accurate-down-to-the-accent one-woman show, "Soaring with Christa McAuliffe" can be seen on YouTube.

John Germ, the former head of Rotary International who proved pivotal in building Chattanooga's Challenger Center, introduced me to June in 2019. I'm pleased to call him a mutual friend. Pam Peterson, Rick Hutto, and Jim Rosebush provided crucial details of the Challenger Centers' early days. Lance Bush hosted a memorable lunch in Washington. Lisa Vernal helped with resources at the Challenger Center headquarters.

Ray Duckler served as my guide to parts of Concord. Former *Monitor* editor Mike Pride, who went on to oversee the Pulitzer Prizes, shared his memories of the year Christa made headlines. He put me in touch with Bob Hohler, whose *"I Touch the Future": The Story of Christa McAuliffe* is a vital account of that year. The *Monitor*'s Ralph Jimenez shared his vivid memories of January 28, 1986. The folks at the Yellow Submarine, the best fish shack in Concord, kept me going on my visits before the pandemic put them out of business. Here's hoping it resurfaces.

Any writer would be lucky to have an editor like Conor Mintzer. I have heard Conor described as a rising star in publishing, but in my book he's already risen. I came to rely on his sharp eye, his judgment, his advocacy, his heart, and his boundless energy. I'm also indebted to Holt's Amy Einhorn, Sarah Crichton, Chris

O'Connell, Maggie Richards, Christopher Sergio, Meryl Levavi, Caitlin O'Shaughnessy, Carolyn O'Keefe, and Pauline Post. At the Robbins Office, my advocate, advisor, ally, and friend David Halpern also serves as my agent. David shaped the project and saw it through to publication. I'm also grateful to Kathy Robbins, Lisa Kessler, Janet Oshiro, and their colleagues at the Robbins Office, and to Paul Golob, who worked with me on the proposal and helped launch the book.

John Logsdon, David Sugar, Brian Holloway, and Joe Dwinnell contributed important details. Julianne Matters at Carnegie Mellon University helped me track down details of Judith Resnik's life and work. I owe nods to Tom and Kelly Cook, Jacqueline Sheehan and Rick Paar, Alexis Johnson and Phil Sullivan, Helen Rosenberg, John Rezek, Jenny Llakmani, Steve Randall, Arthur Kretchmer, Randy Phillips, Steve Sklare, Michael Arkush, and New Jersey's Kubik Circle: Ken Kubik, Chris Carson, and Doug Vogel.

Pamela Marin improves every page of my work. My in-house editor and partner wrote the best book on our shelves, *Motherland*, before turning to screenwriting. The lead writer of *Tommy's Honour*, winner of Scotland's 2016 BAFTA as Best Feature Film, inspires me to try to write better. Thanks to her I also get to thank our children, entrepreneur Calloway Marin Cook, the best golf partner I ever had, and Lily Lady Cook, whose scholarly mind and poet's soul I try to emulate.

After well over a year of research I made contact with Steven McAuliffe, who had not answered reporters' questions about Christa since 1986. I am beyond grateful to Judge McAuliffe.

INDEX

About the Author

Kevin Cook is the author of *Tommy's Honor*, *Titanic Thompson*, *Kitty Genovese*, and *Ten Innings at Wrigley*. He has written for the *New York Times*, *Men's Journal*, *GQ*, *Smithsonian*, and many other publications and has appeared on CNN, NPR, and Fox News. An Indiana native, he lives in Northampton, Massachusetts.